Advances in Termite, Bee and Wasp Biology - Ecology, Physiology, and Integrated Management

Edited by Vonnie D. C. Shields and Robin E. Owen

Published in London, United Kingdom

Advances in Termite, Bee and Wasp Biology - Ecology, Physiology, and Integrated Management
http://dx.doi.org/10.5772/intechopen.1001717
Edited by Vonnie D. C. Shields and Robin E. Owen

Contributors
Akinwande Kayode Lawrence, Ashok Kumar Chaubey, Francine Battaglia, Hiovanis Castillo Pantoja, Isa Gan, Jacques J. M. van Alphen, Javid Bayandor, Kajol Yadav, Lovely Bharti, Maria Luz Alonso, Munjal Shah, Omaira de la Hera, Ricardo Lorenzo Avila Rondon, Rosa Maria Alonso, Sedighe Ashtari, Siapo Yao Martin, Tahiri Annick, Tarik Godoy D. Plaza, Yaimy Marrero Pérez

First published in London, United Kingdom, 2026 by IntechOpen
IntechOpen is the global imprint of INTECHOPEN LIMITED, registered in England and Wales, registration number: 11086078, 167-169 Great Portland Street, London, W1W 5PF, United Kingdom

For EU product safety concerns: IN TECH d.o.o., Prolaz Marije Krucifikse Kozulić 3, 51000 Rijeka, Croatia, info@intechopen.com or visit our website at intechopen.com.

British Library Cataloguing-in-Publication Data
A catalogue record for this book is available from the British Library

Advances in Termite, Bee and Wasp Biology - Ecology, Physiology, and Integrated Management
Edited by Vonnie D. C. Shields and Robin E. Owen
p. cm.
Print ISBN 978-0-85466-242-5
Online ISBN 978-0-85466-241-8
eBook (PDF) ISBN 978-0-85466-243-2

If disposing of this product, please recycle the paper responsibly.

Meet the editors

Vonnie D. C. Shields, Ph.D., is currently a Full Professor in the Biological Sciences Department in the Fisher College of Science and Mathematics, Towson University, USA. Dr. Shields' research explores gustatory, olfactory, and visual cues in insects. Her laboratory employs morphological, behavioral, and electrophysiological techniques to better understand sensory mechanisms by which larval and adult insects find host plants and detect plant-associated volatiles. Another line of research involves identifying olfactory chemicals that elicit walking behavior. Dr. Shields received both a BS and a Ph.D. from the University of Regina, Saskatchewan, Canada. A portion of her Ph.D. studies was conducted at the University of Alberta in Edmonton, Canada. After graduating, she accepted a research associate position to conduct postdoctoral studies at the Arizona Research Laboratories Division of Neurobiology, University of Arizona, USA, before joining the faculty at Towson University, where she rose through the ranks from assistant to full professor. She teaches undergraduate and graduate courses, including anatomy and physiology, histology, and microscopy and histological microtechniques.

Robin E. Owen graduated from the University of Toronto in 1977 with a B. Sc. in zoology and then with a Ph.D. in 1982. After a two-year postdoctoral fellowship at LaTrobe University in Australia, he held a five-year NSERC University Research Fellowship at the University of Calgary. He joined Mount Royal College (now a University) as a full-time instructor in 1989. He is a professor in the Department of Biology and teaches introductory biology, population genetics, invertebrate zoology, and evolutionary biology. His research interests are centered on the evolutionary genetics of the *Hymenoptera* and include: bumble bee genetics, ecology, and taxonomy; theoretical population genetics of X-linked and haplodiploid systems; evolution of eusociality; and insect mimicry.

Contents

Preface

Termites

Termites are eusocial insects in the order Blattodea. They are most often grouped into the infraorder Isoptera with cockroaches. They occur on every continent except Antarctica and are most diverse and abundant in tropical and subtropical regions. Over 3,000 species have been described. They are usually divided into three broad ecological groups based on their nesting and feeding habits: damp-wood, dry-wood, and subterranean termites. Subterranean termites form the largest colonies and are responsible for most of the structural damage to human buildings.

A termite colony is a complex social unit composed of distinct castes: reproductives, workers, and soldiers. The queen may survive for more than ten years in some species and can produce up to several thousand eggs daily. Workers are usually sterile, foraging for food, constructing the nest, caring for their young, and feeding other castes. Soldiers typically have enlarged heads with powerful mandibles or chemical defenses that help safeguard the colony against predators, especially ants. Termites use pheromones and vibrations for communication, enabling coordination of foraging trails, alarm responses, and caste differentiation at the colony level.

Termites are perhaps best known for their ability to digest cellulose, the main structural component of plant cell walls. Most species depend on symbiotic microorganisms in their hindgut (including protists and bacteria) that produce cellulase and other enzymes to degrade wood and plant material. The termite provides a stable, anoxic environment and a constant supply of food for these symbionts. This mutualism is so tightly integrated that many termites must acquire their gut microbes through social behaviors such as trophallaxis, or the exchange of gut contents, after molting or during early development.

Ecologically, termites are major decomposers and ecosystem engineers. They break down dead wood, leaf litter, and other plant debris, accelerating nutrient cycling and influencing soil structure and fertility. The nests and galleries increase soil porosity and can alter water infiltration and gas exchange. In some tropical savannas and forests, termite mounds form nutrient-rich islands that support distinct plant communities and influence local biodiversity patterns. Termites serve as an important food source for many vertebrates and invertebrates, including birds, reptiles, anteaters, and many arthropods.

Termites are both beneficial and problematic. In natural ecosystems, their role in decomposition and soil formation is indispensable. There is growing interest in termite-derived enzymes and gut microbes as models for biofuel production and industrial biomass conversion. At the same time, a small fraction of species are serious pests of wooden structures, plantations, and crops, causing billions of dollars in economic losses worldwide each year. Management strategies often combine chemical barriers or baits with physical exclusion and habitat modification to reduce risk. Current research also explores more targeted, environmentally friendly control methods, including biological control agents and growth regulators, reflecting the broader challenge of balancing termite conservation in natural habitats with protection of human infrastructure.

The chapters on Termites are presented in the first section of the book: "Feeding Patterns of Termites (Isoptera)" by Tarik Godoy D. Plaza, "Termiticidal Activity of Senna occidentalis and Tithonia divercifolia Extracts and Study of Their Toxicity on Mammals" by Annick Tahiri and Yao Martin Siapo; and "Integrated Management of Termites" by Lovely Bharti, Kajol Yadav and Ashok Kumar Chaubey.

Hymenoptera

The Hymenoptera is a highly diverse order of insects containing over 140,000 described, extant species, fewer than the Coleoptera (beetles) and Lepidoptera (moths and butterflies). If undescribed species are included, then the Hymenoptera may be the most speciose of all insect orders. Although the vast majority of the Hymenoptera are solitary species, eusociality has arisen independently multiple times in separate lineages. The truly social, or eusocial, insects are those characterised by a reproductive division of labour with a fertile queen and generally sterile workers. Social insects are among the most ecologically dominant and evolutionarily successful animals on Earth.

Humans have been fascinated by and have had an intimate relationship with the honeybee for millennia. The earliest records date back to the Mesolithic, when cave paintings in the Aran Caves in Spain, at least 8000 years old, show people collecting honey from wild bees high on cliffs. From Greek and Roman times, throughout the Medieval period, and the 16th, 17th, and 18th centuries, understanding of bees - honeybees in particular - ants and wasps gradually increased and became more scientific, culminating in the vast explosion of knowledge in the 19th, 20th, and 21st centuries. However, it is not surprising, given the immense diversity of the Hymenoptera, that there are still many unanswered questions, even some very fundamental ones.

Honeybee colonies reproduce by swarming, where a queen leaves with her retinue of workers. It is the original mother queen (the old queen) who leaves. Jacques van Alphen points out that this is a paradox, as genetically it would make more sense for a new daughter queen to be the one to leave. He suggests that this gives the remaining colony a better chance of surviving infectious diseases. This emphasises the vulnerability of social insect societies to disease, given the high number and density of workers within a colony.

Akinwande Kayode Lawrence discusses this vulnerability as one of the causes of the decline of honeybees and other social insects. There are at least 14 viruses that infect Hymenoptera, as well as multiple other bacterial, fungal, and parasitic pathogens. It is the synergistic interactions among pathogens, together with environmental conditions, including climate change, that make these especially challenging to address.

Parasitoid Hymenoptera are themselves important for biocontrol of other insect pests. Integrated pest management involves these parasitoids as well as the use of pesticides. Sedighe Ashtari discusses management techniques to minimize the side effects of pesticides on *Trichogramma spp*., which are commonly used for the biocontrol of various moth pest species.

It is important to understand the distribution of hymenopteran species and to predict potential changes driven by anthropogenic activities and climate change. Hiovanis Castillo, Yaimy Marrero Perez and Ricardo Lorenzo Avila Rondon developed the statistical tool WASPAS. This uses insect observation data to generate criteria for predicting the abundance of different insect species in urban areas and their changes over time, based on environmental and risk factors. This approach will be very valuable for urban planning and maintaining biodiversity in urban areas.

Although most hymenopteran species are beneficial, others are among the worst invasive species; one is *Vespa velutina nigrithorax*, the Asian or yellow-legged hornet, native to Asia. A single queen was accidentally introduced into Europe in 2004. The wasp subsequently spread at about 80 km/year and now occurs throughout France, over much of Spain, and into other parts of the continent. The biology and spread of this very serious pest species are discussed by de la Hera, Alonso, and Alonso, who present effective control measures to mitigate the detrimental effects of the wasps on honeybee hives in apiaries.

Although human activities have altered the distribution of many species, environmental conditions, such as temperature, are also critical. Knowing the physiological limitations of species is essential for predicting the possible consequences of climate change, for example. Gan, Shah, Bayandor, and Battaglia provide a highly detailed study of the physiological and environmental impact of temperature change on bumblebee flight. They find that increasing ambient temperature decreases lift and aerodynamic power production at both hovering and forward flight. Thus, this is another detrimental consequence of climate change that could impact bumblebee populations.

Vonnie D.C. Shields
Biological Sciences Department,
Fisher College of Science and Mathematics,
Towson University,
Towson, MD, USA

Robin E. Owen
Department of Biology,
Mount Royal University,
Calgary, Canada

Section 1

Termites

Chapter 1

Feeding Patterns of Termites (Isoptera)

Tarik Godoy D. Plaza

Abstract

Termite feeding is ecologically important due to the significant impact these insects have on the environment. As detritivores, termites play a crucial role in decomposing organic matter and forming soil by degrading cellulose and other carbohydrate polymers. They actively participate in nutrient cycling, particularly in ecosystems with woody plants, releasing essential nutrients into the soil. Termites also enhance soil structure and increase its water-holding capacity, contributing to biodiversity maintenance. Termites directly influence soil quality by consuming plant materials, promoting nutrient release, water infiltration, and root development. They contribute to soil particle aggregation, forming stable aggregates that maintain soil structure. The effects on soil structure and biodiversity depend on termite species and soil type. Apart from ecological significance, termites can cause substantial economic and social impacts. Integrated termite management approaches are employed to mitigate these impacts, considering environmental consequences. Understanding termite feeding characteristics is vital for effective management. The specialized digestive tract of termites, along with symbiotic microorganisms, plays a critical role in breaking down complex plant materials. Research on termite feeding can contribute to biofuel production, waste degradation, and ecosystem functioning while aiding in pest management strategies.

Keywords: termites, food patterns, soil, feeding groups, ecology

1. Introduction to termite feeding and the importance of understanding them

Termite feeding is a topic of great importance in the field of ecology and biodiversity since these insects are considered important ecological agents and are capable of significantly modifying the environment in which they inhabit. To understand the importance of termite feeding, it is necessary to understand their characteristics and types of feeding.

Termites are detritivores that have a unique ability to degrade cellulose and other carbohydrate polymers thanks to a symbiosis with bacteria, protozoa, and fungi that inhabit their digestive tract [1]. Therefore, they play an important role in the decomposition of organic matter, especially dead wood, and in soil formation. According to Bignell et al. [2], termite colonies can contain millions of individuals and produce large amounts of excrement and saliva that contain important nutrients for the development of plants and other soil organisms.

IntechOpen

We can say that termites play an important role in nutrient cycling in terrestrial ecosystems, especially in those where vegetation is dominated by woody plants. Feeding on wood by termites, for example, results in the release of essential nutrients, such as nitrogen and phosphorus, into the soil. These nutrients are then made available to other plants and organisms in the ecosystem. In addition, termite activity in the soil can improve its structure and increase its water-holding capacity, contributing to the maintenance of biodiversity.

According to Nunes et al. [3], termite mounds in pastures contribute to nutrient cycling, reduce soil compaction, and increase water infiltration. Constantino [4] state that the presence of termites in Cerrado soils has a positive impact on soil fertility due to the release of nutrients from the decomposition of organic matter present in the termite mound. On the other hand, the removal of termites from forest areas can lead to a decrease in nutrient cycling and alterations in the structure and diversity of the plant and animal community [2].

In addition, feeding on plant materials such as wood directly influences soil quality since their decomposition promotes the release of nutrients and the formation of micro and macro porosities that favor water infiltration and plant root development. Moreover, the presence of termite mounds in the soil can also contribute to the aggregation of particles, forming stable aggregates that help maintain soil structure [5]. Another factor is the ability to modify the dynamics of soil organic matter, such as decomposition rate, stability, and material quality.

The effects of termites on soil structure can be different depending on the termite species and soil type. In sandy soils, termite activity can improve fertility and water retention, while in clay soils, termite activity can promote compaction and decrease porosity [5]. Furthermore, termites can also influence soil biodiversity since the activity of these insects can promote the formation of ecological niches and favor the development of other invertebrate species. Considering that they play a fundamental role in the structuring of ecological communities, we can say that, in turn, they can cause significant economic impacts, especially in the timber industry, where they are responsible for significant losses due to wood consumption. According to Menezes et al. [6], termites are considered the most harmful wood pest in the world. Additionally, they can affect agricultural and livestock production by damaging structures and equipment. In urban areas, termites can cause damage to buildings and furniture.

The social impacts of termites are also relevant. In regions where agriculture is the main source of income, termites can harm production and affect the food security of local populations. In urban areas, the damage caused by termites can be financially significant for low-income populations who do not have the resources to repair their homes. In addition to economic and social impacts, termites also affect the environment. Termite activity can affect soil structure, reducing water and nutrient infiltration capacity, which can lead to soil erosion and decreased biodiversity. Furthermore, termites can influence nutrient cycling, especially in savanna and tropical forest ecosystems.

Management of termites is important to minimize the negative impacts they cause. According to Menezes et al. [6], integrated termite management involves a combination of physical, chemical, and biological methods. It is important to emphasize that termite control should be carried out responsibly and sustainably, minimizing negative effects on the environment. Regarding future perspectives, new technologies, and management methods are being developed for termite control. An example of this is the use of natural products, such as essential oils and plant extracts, which have been effective in termite control [7]. Additionally, studies on the biology and ecology of termites can provide important information for the development of new management methods.

Understanding the feeding characteristics of termites is fundamental for the proper management of these insects in agricultural, forest, and urban environments. For example, the use of treated wood or other materials less susceptible to degradation by termites may be an effective strategy to prevent infestations of these insects in buildings. Furthermore, studying termite feeding can lead to new technologies for biofuel production and degradation of plant waste.

2. Anatomy of digestion

Digestion begins in the mouth, where termites ingest wood and reduce it into progressively smaller particles with the help of their saliva [8]. These particles are then transported to the termite's anterior stomach, where microorganisms digest cellulose through enzymes and other substances [1, 8]. Cellulose is broken down into simple sugars, mainly glucose, through fermentation and hydrolysis processes [1, 8]. The resulting sugars are then absorbed by termites and used as a source of energy [8].

Symbiotic microorganisms play a crucial role in this digestion process by producing enzymes that termites are unable to synthesize [1, 8]. Additionally, the microorganisms also help to protect termites from pathogens and maintain the balance of the digestive tract [8]. In termites, the microbiota can act in five ways in the decomposition of plant material: (1) dissimilatory metabolism of sugar polymers (cellulose, hemicelluloses) and their cleavage into short chains to be used by termites or other symbionts, (2) elimination of oxygen in order to maintain an anoxic environment that allows fermentation in the gut, (3) nitrogen metabolism by fixing atmospheric nitrogen or recycling nitrogen waste, (4) elimination of hydrogen by reducing acetogenesis, (5) degradation of aromatic compounds originating from lignin units (demethylation, deacetylation, decarboxylation and, to some extent, degradation of aromatic nuclei), and (6) humification of complex organic matter and mineralization of carbon from humus proteins in soil-feeders [9].

The anatomy of the termite's digestive tract is highly specialized for the digestion of complex plant materials such as cellulose. Most termites have a digestive tract divided into three main sections: the foregut, midgut, and hindgut [10]. The foregut is the region where carbohydrate breakdown begins and houses microorganisms that produce hydrolytic enzymes capable of breaking down cellulose and other polysaccharides into smaller units [8]. The midgut is responsible for nutrient absorption, while the hindgut is the region where symbiotic bacteria responsible for cellulose degradation reside and where most of the digestion occurs [10].

In termites, the hindgut is composed of five segments: the first proctodeal segment (P1), the enteric valve (P2), the paunch (P3), the colon (P4), and the rectum (P5). P1 is tubular, with variable lengths, and its epithelium forms six longitudinal thickenings. P2, usually funnel-shaped, invaginates into P3, whose internal cuticular armor is developed. Regarding the biological function of P3, it is still uncertain, but much is discussed on the subject. The theories suggest the use of this structure as particle filtering filtering [11, 12], microbial enrichment [13], a defense mechanism for some species, such as Anenteotermes sp., which undergo autolysis [14], as a physical barrier against pathogens, preventing the entry of bacteria and other infectious agents into the digestive system [15], or a combination of these functions.

The digestive tract structure of termites is highly adapted for the efficient breakdown of complex plant materials and is highly dependent on symbiotic microorganisms for cellulose degradation. Identification and characterization of these symbiotic microorganisms are essential for understanding the ecology and functioning of ecosystems in which termites inhabit. The digestive enzymes of termites are produced by symbiotic microorganisms present in the insect's digestive tract and are essential for the breakdown of cellulose and other complex substances present in the termite's diet [16]. The main types of enzymes involved in the digestion process are glycanases, cellulases, hemicellulases, pectinases, proteases, lipases, and phosphatases [17].

Glycanases are responsible for breaking down the polysaccharides present in the termite's diet into simple sugars, facilitating nutrient absorption by the insect [17]. Cellulases, hemicellulases, and pectinases are enzymes responsible for the degradation of cellulose and other plant fibers present in the termite's diet [16]. Proteases and lipases are enzymes responsible for the digestion of proteins and lipids present in the diet [17]. Finally, phosphatases are enzymes that act on the release of phosphate from organic compounds, such as phospholipids, phosphoproteins, and nucleotides [16]. In addition, termite digestive enzymes are highly efficient, allowing the insects to extract the maximum amount of nutrients from their diet [17]. This process is essential for the survival of termites in their natural environment, as the diet of these insects is nutrient-poor and highly complex [16].

Termites have a symbiotic relationship with microorganisms in their digestive tract that are essential for the digestion of cellulose and other wood components that make up a large part of their diet. Most of these microorganisms are bacteria and protozoa that live in the hindgut of termites, where cellulose is fermented and converted into organic acids and other compounds that termites can digest [8]. Protozoa that inhabit the termite digestive tract are also known to provide essential nitrogen and vitamins to their hosts and are capable of digesting chitin present in the cell walls of fungi that make up part of the diet of some termite species [8].

Research has shown that the microbial community present in the termite digestive tract can vary widely between species and even within the same species in different locations. In addition, changes in termite diet can also affect the composition of these microorganisms and, therefore, the efficiency of digestion. Therefore, the presence of these symbiotic microorganisms is fundamental to the survival of termites and their ability to digest cellulose and other wood components. Studies on the ecology of these microbial communities and their function in the termite digestive tract have been a topic of great interest in the field of biology and can provide valuable insights to produce biofuels and other technological applications.

3. Characteristics of termite feeding

Termites are insects that feed on different types of materials, with the main sources of food being wood, soil, and fungi [18]. Wood-based feeding is mainly carried out by subterranean and drywood termites, which can digest the cellulose and lignin present in wood with the help of digestive enzymes produced in their digestive tract [19]. Soil-based feeding is carried out by subterranean termites, which feed on soil particles and organic matter present in that environment [18]. Fungi-based feeding is carried out by mound-building termites, which cultivate fungi in underground gardens and feed on the fruiting bodies produced by these fungi [20, 21]. Lastly, feeding on other materials is less common and includes materials such as paper, fabrics, plastics, and others.

The type of termite feeding can be influenced by various factors, such as availability of food resources, evolutionary adaptation, and symbiosis with microorganisms that aid in digestion and nutrient metabolism [2, 8]. Knowledge of termite feeding is important for their ecological role as decomposers of plant material and nutrient cycling in ecosystems, as well as their economic relevance as urban and forest pests [22].

4. Wood-based feeding

This type of feeding can be carried out in different ways depending on the state of the material that the species has adapted to use, such as through leaf litter and woods in different stages of decomposition (xylophages), live plant material being removed through cutting (foliophages or harvesters), wood in different stages of decomposition in the process of being incorporated with the soil (soil/wood interface), or dry wood, generally used for construction and furniture (drywood).

Cellulose is broken down into smaller molecules by enzymes produced by symbiotic microorganisms present in the anterior gut of the termite [23]. These enzymes include cellulases, hemicellulases, and ligninases, which degrade the main fractions of wood [24]. Cellulose is the main source of energy for termites, while hemicelluloses and lignins are used for the synthesis of secondary compounds, such as volatile fatty acids and phenolic compounds [25].

Termite-mediated decomposition of wood also plays an important role in nutrient cycling in forest ecosystems [26]. In addition, the removal of dead wood and the creation of galleries by termites can influence soil structure and nutrient availability for other plants [27]. However, wood-based feeding can also cause significant economic damage, especially in urban areas, where termite colonies can infest and damage wooden structures [28].

5. Organic matter-based feeding (Humivores)

Termites are social insects that play an important ecological role as decomposers and nutrient recyclers in terrestrial ecosystems. Their diet is based on different sources of organic matter, including organic matter present in the soil. This consumption by termites can vary according to the species and the geographic region where they are present [21].

Termites that consume organic matter usually do so in order to obtain nutrients and water to survive and for the construction of their nests [21]. Soil-based feeding can be carried out in different ways by termites, such as direct ingestion of soil particles or through the collection of plant material contaminated with soil [29].

The ability of termites to consume soil and modify its chemical and physical characteristics is widely recognized [21]. Their soil-based diet can also affect soil structuring and nutrient cycling in terrestrial ecosystems [30].

Some studies highlight the importance of organic matter-based feeding for maintaining ecosystem health, such as regulating the nutrient cycle and preserving soil invertebrate biodiversity [31]. However, in some situations, such as in agricultural areas, soil-based feeding by termites can have a negative impact, causing economic losses.

Management of organic matter-based feeding by termites can be carried out through different strategies, such as the use of physical or chemical barriers to prevent termites' access to crop areas [29]. In addition, studies are being conducted to evaluate the use of microorganisms and plant extracts as alternatives to chemical control of termites in agricultural areas.

In summary, soil-based feeding by termites is a relevant topic in studies related to the ecology of terrestrial ecosystems, with significant impacts on soil structuring and nutrient cycling. Management of this feeding can be carried out through different strategies, with the potential to contribute to the preservation of soil invertebrate biodiversity and the reduction of economic impact caused by termites in agricultural areas.

6. Feeding tenant

It is not yet clear what the feeding of tenant species is, and there are some cases of specific feedings, such as *Serritermes serrifer,* which possibly feeds on plant organic material inside the nests of other species, since the foraging tunnels of *S. serrifer* follow high concentrations of this material [32].

Another case is *Inquilitermes* fur, which, due to the microorganisms present in its intestine, possibly feeds on the feces of Constrictotermes cyphergaster, which in turn feeds on fungi [33]. However, we emphasize that both examples mentioned are tenant species of other species, and there are still few studies on their feeding, which makes it necessary to further investigate this issue, considering that both studies are inconclusive regarding feeding.

7. Fungi and lichen-based diet

The fungi and lichen-based diet is carried out by termites of the *Macrotermitinae* subfamily, which cultivate fungi in special chambers within the nest. The termites provide the substrate for the fungus to grow, while the fungus digests cellulose to produce metabolites that will be used as food by the termites [20].

The relationship between termites and fungi is considered a mutualistic symbiosis, as both organisms benefit from the relationship. Moreover, this symbiosis is considered one of the most efficient in nature, with a food efficiency rate of up to 90% [34]. The fungi cultivated by termites also play an important role in soil structure. Through the production of exoenzymes, these fungi degrade organic matter in the soil, releasing important nutrients for nutrient cycling and soil fertility maintenance [21].

In addition, the fungi-based diet can also have important economic impacts. Some termite species cultivate edible fungi, which are used in human consumption in some traditional cultures, such as the *Termitomyces* genus in Africa [35].

Regarding the management of termite diets based on fungi, a possible strategy would be the use of fungicides to prevent the growth of fungi cultivated by termites. However, this strategy can have negative effects on soil fertility and nutrient cycling [21]. Finally, it is important to highlight that there is still much to be studied regarding the termite diet based on fungi, especially regarding the diversity of fungi cultivated by different termite species and the function of these fungi in maintaining ecosystem health [21].

8. Conclusions

Although cellulose is the only source of food resources for termites, this substance is consumed in various forms found in nature, such as live and dead wood in different stages of decomposition, grasses, fungi, lichens, dry leaves, and humus [36]. In general, because they are detritivores, termites use a virtually unlimited amount of food resources since they are not constrained by the availability of the resource and do not harm the speed of resource recovery in nature, as herbivores can exhaust resources.

There are termite species with a broad diet, especially wood-feeders, which are responsible for most of the damage to wood structures and agriculture, such as some species of Termitidae (*Nasutitermes*), *Kalotermitidae (Cryptotermes)*, and *Rhinotermitidae (Coptotermes)*, or species considered pests due to their damage in urban environments, such as *Kalotermitidae (Cryptotermes brevis, Cryptotermes dudleyi, Cryptotermes havilandi)*, and *Rhinotermitidae* (*Coptotermes gestroi*) [4]. For monoculture plantations such as sugarcane, rice, and eucalyptus, we have pest species from the families Termitidae (*Nasutitermes, Cornitermes, Procornitermes, Cylindrotermes, and Neocapritermes) and Rhinotermitidae (Heterotermes*) [37–39].

According to Eggleton et al. [40], termites can be classified based on their feeding type, with soil-feeders (mineral soils) and wood-feeders, which feed on litter and lichens. Another classification followed the pattern: wood-feeders, humus-feeders, intermediates (wood-humus interface), leaf-feeders, some are fungus-cultivators, the Macrotermitinae, and those that feed on mosses and lichens [41, 42]. We must highlight that fungus cultivators are common in the Paleotropics but are not found in the Neotropical and Australasian regions [41, 42]. In this sense, several researchers have attempted to categorize termites into different ecological groups based on the type of food consumed [40–42], related food to nest morphology [43, 44], food to morphological characters and gut contents [11], various environmental factors, and food [26].

In 2002, Bignell and Eggleton proposed six functional groups of termites based on their feeding habits: Soil-feeders, termites that feed on soil; Wood-feeders, those that feed on wood and litter, regardless of its state of decomposition; Soil/wood interface-feeders, those that consume extremely rotten wood, almost merging with the soil; Litter-foragers, those that feed on pieces of wood in the litter; Grass-feeders, those that usually feed on dead or dry grass; and Minor feeding groups, represented by a small group of species that consume lichens and algae.

Donovan et al. [11] suggested a classification of food groups based on morphological characters and gut contents of animals in the termite collection at the Natural History Museum in London. They obtained four distinct groups: Group I, with "lower termites" (possessing flagellated microorganisms) feeding on dead wood and grass; Group II, part of the family Termitidae, with a wide variety of food sources, such as rotten wood, grass, litter, and microepiphytes; Group III, some representatives of the Termitidae family feeding on soil layers rich in organic material; Group IV, containing termites that feed on soil (again some Termitidae), apparently ingesting mineral soil.

9. Applications

The study of termite feeding is an area that has been gaining importance in recent years due to its implications in ecology, agriculture, engineering, and medicine. Due to their ability to decompose cellulose and lignin, termites have potential for use in the production of biofuels and in the bioremediation of contaminated soils. In addition, termites are known to present a great diversity of microorganisms in their digestive tract, which play a fundamental role in the digestion of cellulose and lignin. These microorganisms have attracted the attention of researchers interested in developing new biocatalysts and bioprocesses for the production of bioenergy and chemicals from biomass [45]. Another important aspect is the relationship between termite feeding and soil structure. Termites are capable of altering the physical and chemical characteristics of soil, contributing to the formation of microhabitats and increasing soil fertility. Studies in this area are important for the development of sustainable land management practices and biodiversity conservation.

However, despite recent advances, there is still much to be discovered about termite feeding and its implications in ecology and agriculture. New techniques of genetic sequencing and metagenomic analysis can help to elucidate the diversity of microorganisms present in the termite gut and their functions in the digestion of cellulose and lignin. In addition, comparative studies between different termite species can help to understand the evolutionary adaptations that have allowed these insects to successfully colonize a variety of habitats and ecosystems [45].

Finally, it is important to highlight that research on termite feeding is not limited only to biological and ecological aspects. There are also important social and economic implications, such as the application of termite control techniques in urban and rural areas, the development of technologies for the production of biofuels, and the generation of employment in industries related to bioenergy and bioremediation.

In summary, research on termite feeding is an area in constant evolution with broad implications in various fields of knowledge. New discoveries and advances in this area can contribute to the development of more sustainable and efficient solutions for the production of food, energy, and materials, as well as helping to preserve biodiversity and ecosystem services.

Author details

Tarik Godoy D. Plaza
Zoology Museum of the University of São Paulo, São Paulo, Brazil

*Address all correspondence to: tarikplaza@hotmail.com

References

[1] Scharf ME, Karl ZJ, Sethi A, Boucias DG. Multiple levels of synergistic collaboration in termite lignocellulose digestion. PLoS One. 2011;**6**(7):e21709. DOI: 10.1371/journal.pone.0021709 Epub 2011 Jul 1

[2] Bignell DE, Roisin Y, Lo N. Biology of Termites: A Modern Synthesis. Dordrecht: Springer; 2011. DOI: 10.1007/978-90-481-3977-4

[3] Nunes LAPL, Santos VB, Tosta MS, Santos CAF, Souza TT. Termites promote nutrient cycling and water infiltration in an Amazonian pasture. Applied Soil Ecology. 2019;**144**:59-66

[4] Constantino R. The pest termites of South America: Taxonomy, distribution and status. Journal of Applied Entomology. 2002;**144**(7-8):521-537

[5] Funari CS et al. High resolution ultra high pressure liquid chromatography–time-of-flight mass spectrometry dereplication strategy for the metabolite profiling of Brazilian Lippia species. Journal of Chromatography A. 2012;**1259**:167-178

[6] Menezes EB, Aquino AM, Menezes ELA, Correia MEF, Souza JH, Souza RM. Cupins: taxonomia, biologia, ecologia e sua importância nos sistemas agropecuários. Embrapa Agrobiologia. 2007. 53 p

[7] Zhu Y, Li C, Tao X, Brazill JM, Park J, Diaz-Perez Z, et al. Nmnat restores neuronal integrity by neutralizing mutant Huntingtin aggregate-induced progressive toxicity. Proceedings of the National Academy of Sciences of the United States of America. 2019;**116**(38):19165-19175

[8] Brune A. Symbiotic digestion of lignocellulose in termite guts. Nature Reviews Microbiology. 2014;**12**(3): 168-180. DOI: 10.1038/nrmicro3182

[9] Šobotník J, Dahsjo CAL. Isoptera. In: Module in Life Sciences. 2017; Elsevier. ISBN: 978-0-12-809633-8. DOI: 10.1016/B978-0-12-809633-8.02256-1

[10] Brune A, Dietrich C, Moissl-Eichinger C. Role of the termite gut microbiota in symbiotic digestion. In: Breitenbach J, Schmidt O, Paulick S, editors. Frontiers in Ecology, Evolution and Complexity. Cham: Springer; 2015. pp. 219-232

[11] Donovan SE, Eggleton P, Bignell DE. Gut content analysis and a new feeding group classification of termites. Ecological Entomology. 2001;**26**(4):356-366

[12] Bignell DE, Roisin Y, Lo N. Biology of Termites: A Modern Synthesis. Dordrecht: Springer; 2011. DOI: 10.1007/978-90-481-3977-4

[13] Scheffrahn RH. Compositermes vindai (Isoptera: Termitidae: Apicotermitinae), a new genus and species of soldierless termite from the Neotropics. Zootaxa. 2013;**3652**(3):381-391

[14] Sands WA. The soldierless termites of Africa (Isoptera: Termitidae). Bulletin of the British Museum (Natural History), Entomology Supplement. 1972;**18**:1-244, pls. 1-9

[15] Nalepa CA. Origin of termite eusociality: Trophallaxis integrates the social, nutritional, and microbial environments. Ecological Entomology. 2015;**40**:323-335. DOI: 10.1111/een.12197

[16] Martinez D, Berka RM, Henrissat B, et al. Genome sequencing and analysis of the biomass-degrading fungus Trichoderma reesei (syn. Hypocrea jecorina). Nature Biotechnology. 2008;**26**(5):553-560

[17] Wang Y, Li H, Li D, Li X, Liang Q, Chen Y, et al. Characterization and phylogenetic analysis of cellulase genes in the beetle Tenebrio molitor. International Journal of Biological Sciences. 2013;**9**(3):328-339

[18] Constantino R. Catalog of the termites (Isoptera) of the Neotropical region. Neotropical Entomology. 2019;**48**(1):1-165

[19] Cleveland LR, Hall SR, Sanders EP, Collier J. The wood-feeding roach Cryptocercus, its protozoa, and the symbiosis between protozoa and roach. Memoirs of the American Entomological Society. 2011;**4**:1-369

[20] Aanen DK, Eggleton P, Rouland-Lefèvre C, Guldberg-Frøslev T, Rosendahl S, Boomsma JJ. The evolution of fungus-growing termites and their mutualistic fungal symbionts. Proceedings of the National Academy of Sciences. 2002;**99**(23):14887-14892

[21] Bignell DE, Eggleton P, Nunes L. Termites as mediators of soil structure in the tropics: Causes and effects. Geoderma. 1997;**79**(1-4):229-250

[22] Su NY, Scheffrahn RH. Intra- and interspecific competition of the Formosan and eastern subterranean termite: Evidence from field observations (Isoptera: Rhinotermitidae). Sociobiology. 1988;**14**:157-164

[23] Breznak JA, Brune A. Role of microorganisms in the digestion of lignocellulose by termites. Annual Review of Entomology. 1994;**39**(1):453-487

[24] Lima NCR, Bochatey EL, Sant'Anna GL Jr. Digestão em cupins: mecanismos e potencial biotecnológico. Química Nova. 2015;**38**(6):791-800

[25] Bignell D. Towards a universal sampling protocol for soil biotas in the humid tropics. Pesquisa Agropecuaria Brasileira - PAB. 2009;**44**. DOI: 10.1590/S0100-204X2009000800005

[26] Santos MN, Teixeira MLF, Pereira MB, Menezes EB. Environmental factors influencing the foraging and feeding behavior of two termite species (Isoptera: Rhinotermitidae) in natural habitats. Sociobiology. 2010;**55**(3):763-777

[27] Lavelle P, Decaëns T, Aubert M, Barot S, Blouin M, Bureau F, et al. Soil invertebrates and ecosystem services. European Journal of Soil Biology. 1995;**31**(4):199-215

[28] Husseneder C. Termites as pests of buildings. Insects. 2000;**1**(2): 76-94

[29] Chouvenc T, Li HF, Austin J, Bordereau C, Bourguignon T, Cameron SL, et al. Revisiting Coptotermes (Isoptera: Rhinotermitidae): A global taxonomic road map for species validity and distribution of an economically important subterranean termite genus. Systematic Entomology. 2016;**41**:299-306. DOI: 10.1111/syen.12157

[30] Jouquet P, Bottinelli N, Shanbhag RR, Bourguignon T, Traoré S, Abbasi SA. Termites: The neglected soil engineers of tropical soils. Soil Science. 2016;**181**(3/4):157-165

[31] Boucias DG, Cai Y, Sun Y, Lietze V-U, Sen R, Raychoudhury R, et al. The hindgut lumen prokaryotic microbiota of the termite Reticulitermes flavipes and its responses to dietary lignocellulose

composition. Molecular Ecology. 2013;**22**:1836-1853

[32] Barbosa JRC, Constantino R. Polymorphism in the neotropical termite Serritermes serrifer. Entomologia Experimentalis et Applicata. 2017;**163**:43-50

[33] Barbosa-Silva AM, Faria MAA, Mello AP, Souza AEF, Garcia HHM, Bezerra-Gusmão MA. Lignocellulosic fungi in nests and food content of Constrictotermes cyphergaster and Inquilinitermes fur (Isoptera, Termitidae) from the semiarid region of Brazil. Fungal Ecology. 2016;**20**:75-78

[34] De Fine Licht HH, et al. Laccase detoxification mediates the nutritional alliance between leaf-cutting ants and fungus-garden symbionts. In: Proceedings of the National Academy of Sciences of the United States of America. 2013;**110**(2):583-587

[35] Nyeko P, Stewart J, Franzel S, Barklund P. Farmer's experiences in the management and utilization of Calliandra Calothyrsus, a fodder shrub, in Uganda. AgREN, Network Paper No 140. 2004. p. 14

[36] Eggleton P, Bignell DE. Monitoring the response of tropical insects to changes in the environment: Troubles with termites. In: Harrington R, Stork NE, editors. Insects in a Changing Environment. London: Academic Press; 1995. pp. 473-497

[37] Costa-Leonardo AM. Digestive physiology of the termites (Isoptera). Brazilian Journal of Biology. 2002;**62**(4A):609-622

[38] Novaretti WRT, Fontes LR. Cupins: Uma grave ameaça à cana-de-açúcar no Nordeste do Brasil. In: Fontes LR, Berti Filho E, editors. Cupins: O desafio do conhecimento. Piracicaba: FEALQ; 1998. pp. 163-172

[39] Milano S, Fontes LR. Cupim e cidade. Implicações ecológicas e controle. São Paulo; 2002. p. 141

[40] Eggleton P, Homathevi R, Jeeva D, Jones DT, Davies RG, Maryati M. The species richness and composition of termites (Isoptera) in primary and regenerating lowland dipterocarp forest in Sabah, East Malaysia. Ecotropica. 1997;**3**:119-128

[41] Bignell DE, Eggleton P. Termites in ecosystems. In: Termites: Evolution, Sociality, Symbioses, Ecology. Dordrecht: Springer; 2000a. pp. 363-387

[42] Bignell DE, Eggleton P. Termites in ecosystems. In: Abe T, Higashi M, Bignell DE, editors. Termites: Evolution, Sociality, Symbiosis, Ecology. Dordrecht: Kluwer Academic Press; 2000b

[43] Domingos DJ, Cavenagui TMC, Gontijo MA, Drumond MA, Carvalho KCF. Composição em espécies, densidade e aspectos biológicos da fauna de térmitas de cerrado em Sete Lagoas, MG. Ciência e Cultura, São Paulo. 1986;**38**(1):199-207

[44] Gontijo TA, Domingos DJ. Guild distribution of some termites from cerrado vegetation in Southeast Brazil. Journal of Tropical Ecology. 1991;7:523-529

[45] Brune A. Methanogenesis in the digestive tracts of insects and other arthropods. In: Biogenesis of Hydrocarbons. Handbook of Hydrocarbon and Lipid Microbiology. Cham: Springer International Publishing; 2018. pp. 1-32

Chapter 2

Termiticidal Activity of *Senna occidentalis* and *Tithonia divercifolia* Extracts and Study of Their Toxicity on Mammals

Siapo Yao Martin and Tahiri Annick

Abstract

The termiticidal activity of methanolic extracts of *Senna occidentalis* and *Tithonia diversifolia* leaves was determined on the crop pest termite, *Ancistrotermes cavithorax*. Both extracts were toxic to termite workers by contact and inhalation. The *T. diversifolia* extract (LC50 of 29, 76 mg/l) was more toxic than the *S. occidentalis* extract (LC50 of 84.90 mg/l). The two extracts are not repellent but anti-palatable. To recommend insecticidal plants in the culture medium requires the knowledge of their harmfulness to the environment and mammals. This and the toxicity of its two extracts were determined on female rats *Rattus norvegicus*. The LD 50 of these two extracts exceeded 5000 mg/kg of body mass. The extracts did not cause any renal or hepatic damage after 14 days. The use of these insecticidal plant extracts can therefore be recommended to farmers.

Keywords: toxicity, termites, mammals, *Tithonia diversifolia*, *Senna occidentalis*

1. Introduction

Termites are one of the greatest scourges in tropical agriculture and agroforestry [1, 2]. In Côte d'Ivoire, West Africa, they regularly attack and destroy food and industrial crops [3–6]. To control termite attacks and damage, several methods are used by producers. Among these techniques is the use of chemical inputs [7]. Excessive use of chemical inputs can adversely affect living organisms and their environment [8] with very high repair costs. In sub-Saharan Africa, the potential cost of treating pesticide-related illnesses between 2005 and 2020 is about US$90 billion [9]. These problems with synthetic pesticides have led to the search for new termiticidal substances from plants that are environmentally and human health friendly [5, 10, 11]. Pesticidal plants are toxic to pests and generally less harmful to beneficial insects and the environment [12]. The present study set out to evaluate the insecticidal properties of two common plants in West Africa, *S. occidentalis* and *T. diversifolia*, on the crop pest termite, *Ancistrotermes cavithorax*. The toxicity of these two plants will also be evaluated on the rat *Rattus norvegicus*.

2. Materials and methods

2.1 Plant materials

Fresh leaves of *S. occidentalis* and *T. diversifolia* were collected for this study in May 2017 in the commune of Bingerville (Abidjan – Ivory Coast) (between 6 and 7 am). The harvested samples were cleaned, air-dried at room temperature, and protected from light for 2 weeks. They were then looted and well-preserved until use.

2.2 Animal material

The pesticide tests were conducted on adult workers of the termite *Ancistrotermes cavithorax* captured in underground nests in the commune of Bingerville (Abidjan – Ivory Coast).

The acute toxicity tests and the histopathology study were conducted on *R. norvegicus* rats of the Wistar strain from the animal house of the Ecole Normale Supérieure d'Abidjan (ENS). These rats are virgin, nulliparous, nonpregnant, 6–8 weeks old, and between 140 and 160 g. The rats were exposed to 5 days of acclimatization with free access to water and food *ad libitum*. The bedding was renewed twice a week.

2.3 Preparation of extracts

Extracts were prepared according to the method described by Kaushik and Vir [13]. Thus, 30 g of powder from the leaves of each plant was mixed in 1 l of 70% methanol. The resulting mixture was stirred for 48 h at room temperature (25°C) using a magnetic stirrer type IKAMAG RCT (Staufen, Germany). The mixture was filtered three times on Whatman paper (No. 4) and then using cotton. The collected solution was subjected to rotary evaporation with a rotavapor BUCHI R-100/32 to obtain the methanolic extract. This solution was then dried under vacuum.

2.4 Biological tests

2.4.1 Toxicity by contact

The test was carried out in Petri dishes 90 mm in diameter containing 3.5 g of soil moistened with 1 ml of distilled water [14]. Using a micropipette, 10, 20, 50, and 100 μl doses of each extract were deposited and mixed with the soil. After depositing, the dishes are air-dried for 1 h. Fifty adult workers of *A. cavithorax* termites were introduced into each of these Petri dishes. Each extract is tested at four doses 10, 20, 50 and 100 μl. Each dose was repeated six times. Each control box was treated with the solvent (70% methanol + distilled water). Dead workers were counted every 2 h. After 24 h, the count was carried out every day until all insects were dead. Percentage mortality (PM) is calculated as the ratio of the observed number of deaths to the total number of insects:

$$PM = \frac{Observed\ mortality}{Total\ number\ of\ insects} \times 100 \text{ Total number of insects } = 50/\text{box}. \quad (1)$$

LD50 is calculated on the basis of 24-h mortality.

2.4.2 Inhalation toxicity

The inhalation test was performed in a large Plexiglass® box of 180 × 20 × 70 mm height according to the method of Tahiri et al. [14]. Into each large box was introduced a Petri dish containing 3.5 g of soil previously moistened with 1 ml of distilled water. Fifty *A. cavithorax* workers are introduced into the Petri dishes, which are then closed with a mosquito net lid. Using a micropipette, the 100 µl dose is placed on 4 cm^2 Whatman® paper which is then transferred to the large dish. The device was then hermetically sealed. Each tested solution was repeated six times for all the tests. Each control box was treated with the corresponding solvent. Worker mortality was determined after 24 h of treatment.

2.4.3 Barrier soil test

The barrier soil test was conducted according to the method of Tahiri [11]. Nine plastic boxes (chambers) (diameter 5 × 3.5 cm high), closed with plastic covers, were used per arena. One introduction chamber containing sieved nest soil, four substrate chambers containing nest soil, and four food chambers, each containing two pieces of Whatman no. 4 paper of 4 cm^2 were used.

The chambers were connected to each other and to the introduction chamber located in the center of the arena by transparent tubes (7 cm length) (LABELL Ch/Fr), length 49 inches. In each trial, only one substrate chamber per arena was treated as barrier soil (50 g of sand moistened with 7.5 ml of the product) to prevent termites from reaching the food source. The other untreated substrate chambers contained 50 g of sand moistened with 7.5 ml of distilled water.

Three hundred *A. cavithorax* workers are weighed before being introduced into the arena through the introduction chamber. They remained isolated for 24 h in this chamber (clamped connecting tubes) to acclimatize before moving freely through all chambers of the arena. The arena is maintained in a room at 27°C. Three replicates with 300 termites were performed for each trial with the extracts and distillates of *S. occidentalis* and *T. diversifolia* and for each corresponding control.

After 10 days of continuous exposure to the product, surviving workers are counted in all arenas, and their total weight is calculated. The tunneling task and food consumption of workers were compared between treated and untreated insects.

2.4.4 Acute toxicity in female rats

The experiment was conducted according to the European OECD guideline 423 [15]. Nine female rats were divided into three batches of three rats (one control and two treated batch) and used for the experiments. The animals were deprived of food overnight. Prior to the experiment, the animals were provided free access to water. Control animals received ml/100 g body weight of distilled water. Due to the low toxicity of *S. occidentalis* [16] and *T. diversifolia* [17] leaves, the treated batch received a single dose of 2000 mg/kg body weight of the treatment *T. diversifolia* or *S. occidentalis*. Each animal received 1 ml/100 g body weight of either *T. diversifolia* or *S. occidentalis* extract. The change in general behavior or mortality of rats in each batch was monitored for 14 days. After this time, the animals were introduced into boxes containing cotton soaked in ether in order to anesthetize them. After anesthesia, the animals were sacrificed.

2.4.4.1 Relative mass of vital organs

At the end of the 14-day test, each animal was weighed and then sacrificed. The kidneys, liver, and heart were removed and weighed. Relative masses were calculated according to the following formula:

$$\text{Relative mass } (\%) = \frac{\text{Mass of the organ}}{\text{mass of the animal}} \times 100 \tag{2}$$

2.4.4.2 Hematologic and biochemical examinations

At the time of animal sacrifice, volumes of blood were collected in EDTA tubes and dry tubes for hematologic and biochemical analysis, respectively. The hematologic analysis was performed using a Sysmex XS-500i automated analyzer. The blood count consisted of determining the number of white blood cells (WBC), red blood cells (RBC), and blood platelets (PLT), the hematocrit (HCT) and hemoglobin (HGB), the mean blood volume (MCV), the mean corpuscular hemoglobin (MCH) and the mean corpuscular hemoglobin concentration (MCHC).

Biochemical analysis of blood was performed after centrifugation at 3000 rpm for 4 min. An aliquot of serum was collected in Eppendorf tubes. Urea, creatinine, blood glucose, cholesterol, alanine aminotransferase (ASAT), and aspartate aminotransferase (ALAT) were determined using a HITACHI 704 R (Japan) automated analyzer.

2.4.4.3 Histopathology of the liver, heart, and kidney

During the sacrifice, the liver, heart, and kidney of treated and control rats were removed, rinsed with 9‰ NaCl solution, and then fixed in 10% formalin. Flaps of each organ were arranged in cassettes and successively bathed in 80°, 90°, 96°, and 96° ethanol for dehydration. They were then cleared n toluene I and II for 1, 2, and 2 h and then cleared in a step for 2 h in the same reagent. The cassettes were removed from toluene, drained, and impregnated, respectively, in two liquid kerosene baths (I and II) for 2 and 3 h in an oven at 50°C. The cassettes were then cured in the open air and then in the freezer. Sections of 5 μm were made with a Leica RM 2125 RTS® microtome and then stained with hematoxylin and eosin (H&E). The sections were observed with an Olympus CKX41 microscope (Germany) connected to a computer equipped with Videomet software. Pictures of the different histological sections were taken.

2.5 Statistical analysis

The results were expressed as mean plus or minus standard deviation, compared by the ANOVA test and separated by the Newman–Keuls test at the 5% threshold. The statistical software used was Statistica 7.0.

3. Results

3.1 Toxic effect by contact of the extracts

The results of the contact toxicity test indicate that the methanolic extract of *S. occidentalis* causes 22.8% mortality after 24 h and 100% after 96 h at the concentration of 10 mg/l, while the control shows 2.4% mortality after 24 h and 100% after

144 h. Concentrations of 1, 2, and 5 mg/l cause less than 20% mortality after 24 h and 100% after 120 h. Termite mortality increased proportionally with concentration and exposure time. Fifty percent (50%) of the mortality of *A. cavithorax* workers was obtained before 48 h of treatment at the 10 mg/l concentration (**Figure 1**). The Kruskal–Wallis test indicated that the total methanolic extract of *S. occidentalis* caused a significantly different mean number of mortalities than the control at 24 h ($P < 0.05$), 48 h ($P < 0.05$) and 72 h ($P < 0.05$).

Percentage mortality as a function of time for the contact toxicity test of the methanolic extract of *Tithonia diversifolia* showed that the 10 mg/l concentration causes 30.8% mortality after 24 h and 100% after 120 h. Concentrations 1, 2, and 5 mg/l caused less than 15% mortality after 24 h and 100% after 120 h, while the control showed 1.2% mortality after 24 h and 100% after 144 h.

The mortality rate of termites increased proportionally with the concentration and time of exposure. Fifty percent (50%) of the mortality of *A. cavithorax* workers is obtained before 48 h of treatment with the 10 mg/l concentration (**Figure 2**). The Kruskal–Wallis test indicated that the extract of *T. diversifolia* tested caused a significantly different average number of mortality than those caused in the control in 24 h ($P < 0.05$), in 48 h ($P < 0.05$), in 72 h ($P < 0.05$) and in 96 h ($P < 0.05$).

3.2 Lethal concentrations of extracts

The Probit analysis performed on the basis of 24-h mortality, showed that the amount of the methanolic extract of *T. diversifolia* required to kill 50% of the tested termite population, is 297.657 µl, or a lethal concentration 50 (LC50) of 29.76 mg/l. With the methanolic extract of *S. occidentalis*, the dose required to kill 50% of the termite population tested, is 849.068 µl, or a lethal concentration 50 (LC50) of 84.90 mg/l of extract (**Table 1**).

3.3 Inhalation toxicity of extracts

After 24 h of inhalation, the mortality rates of 95.99% and 91.66% obtained with the methanolic extract of *T. diversifolia* and *S. occidentalis* leaves, respectively, were

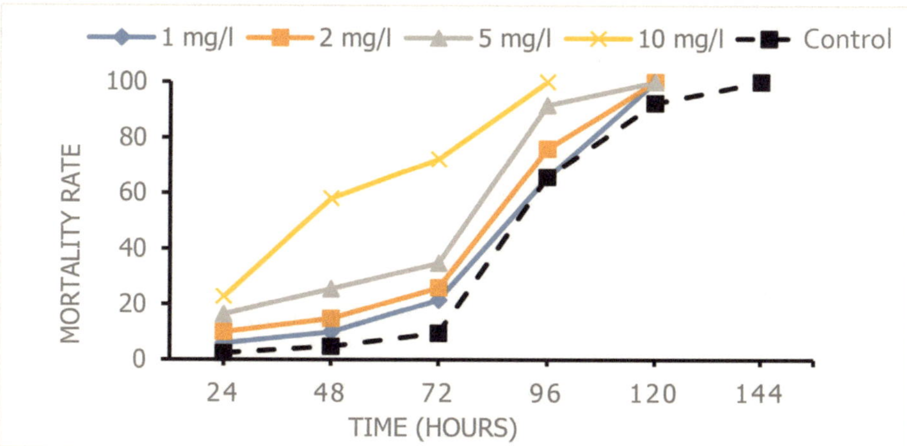

Figure 1.
Variation in mortality of Ancistrotermes cavithorax *after exposure to methanolic extract of* S. occidentalis.

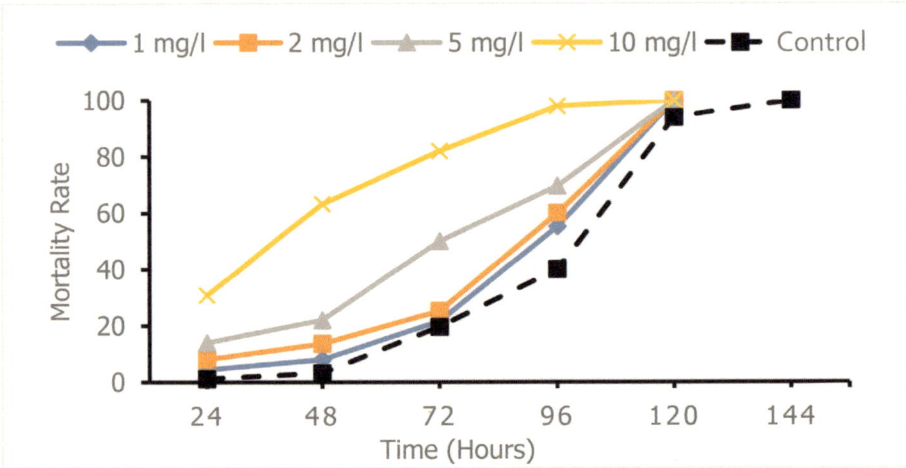

Figure 2.
Variation in mortality of Ancistrotermes cavithorax *after exposure to methanolic extract of* Tithonia diversifolia.

Products	LC50 (mg/l)	Boundary below 95% (mg/l)	Borne above 95% (mg/l)
T. diversifolia extract	29.76	19.87	55.90
S. occidentalis extract	84.90	37.57	426.22

Table 1.
Lethal concentration 50 (LC50) of the total methanolic extract of S. occidentalis *and* Tithonia diversifolia.

higher than that obtained in the control. Statistical analysis indicated that the mortality rate obtained with the total methanolic extract of *T. diversifolia* and *S. occidentalis* was significantly higher ($P < 0.05$) than that of the control (**Figure 3**).

3.4 Toxic effect by barrier soil of extracts

In the test and control arenas, after 10 days of continuous exposure of 300 workers to moistened soil, the mortality rates of 88.89 ± 3.37% and 85.33 ± 9.61%, respectively, in the tests with *T. diversifolia* extracts with and *S. occidentalis* were significantly higher than that obtained in the controls ($P < 0.05$) (**Table 2**). Mortality was unevenly distributed across the four substrate and food chambers in both trials and controls. In the test arenas, the substrate chamber treated with methanolic extracts of both extracts is visited by termites. However, no paper (food bait) was located after the treated substrate chamber was consumed. In contrast, papers in the other untreated substrate chambers of the arena were consumed, but with significantly lower consumption ($P < 0.05$) than in the control areas (**Table 3**).

3.5 Mortality and symptoms in female rats

Single-dose oral administration of 2000 mg/kg body weight of methanolic extract of *T. diversifolia* and *S. occidentalis* to female rats resulted in no deaths after 14 days. Furthermore, observation of the animals after 30 min, 4 and 24 h, and then regularly

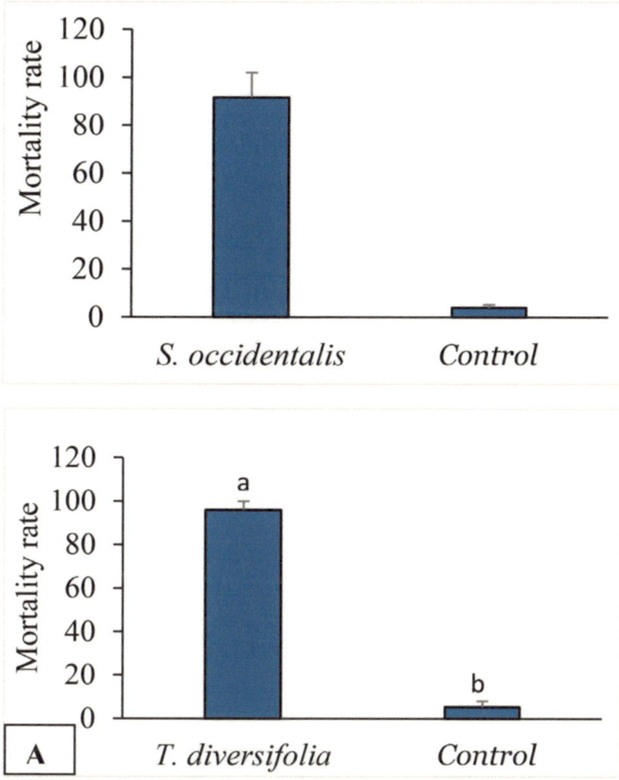

Figure 3.
Distribution of mortality of Ancistrotermes cavithorax *workers during the inhalation test. A: Inhalation test with methanolic extract of* Tithonia diversifolia. *B: Inhalation test with methanolic extract of* S. occidentalis.

Soil moistening	Average mortality (%) ± standard deviation
Extract of *S. occidentalis*	85.33 ± 9.61 a
Extract of *Tithonia diversifolia*	88.89 ± 3.37 a
Distilled water	26.0 ± 1.85 c

Values followed by the same letters in the same column are not significantly different at the 5% threshold (Kruskal–Wallis). N = 300 workers/test; paper area = 2 × 4 cm^2/room.

Table 2.
Average mortality rate of Ancistrotermes cavithorax *workers after 10 days during the barrier soil test.*

every 24 h for 14 days showed no signs of toxicity (salivation, drowsiness, morbidity, and coma). The animals showed signs of well-being (active movement, normal intake of food and water) compared to controls (**Table 4**).

3.6 Effects of extracts on relative organ mass

There was no significant change in the relative mass of the kidneys, liver, and heart of treated rats compared with the relative mass of the organs of control rats (ANOVA, $P > 0.05$) (**Table 5**).

Soil moistening	Cumulative consumption (mm^2) ± standard deviation
Extract of *S. occidentalis*	08.04 ± 1.82 a
Extract of *Tithonia diversifolia*	06.09 ± 1.5 a
Distilled water	175.92 ± 8.46 c

Values followed by the same letters in the same column are not significantly different at the 5% threshold (Kruskal–Wallis). N = 300 workers/test; paper area = 2 × 4 cm^2/room.

Table 3.
Cumulative consumption of Ancistrotermes cavithorax *workers after 10 days during the barrier soil test.*

Rats	Doses (mg/kg)	Number of dead rats	Symptoms
Lot 1 (Control, n = 3)	0	0	—
Lot 2 (EMSO, n = 3)	2000	0	—
Lot 3 (EMTD, n = 3)	2000	0	—

EMSO, Methanolic extract of S. occidentalis; *EMTD, Methanolic extract of* T. diversifolia.

Table 4.
Acute toxicity to female rats of methanolic extracts Tithonia diversifolia *and* S. occidentalis.

Parameters	Lot 1 control	Lot 2 *S. occidentalis*	Lot 2 *T. diversifolia*
Kidneys	0.256 ± 0.005 a	0.268 ± 0.006 a	0.256 ± 0.005 a
Livers	3620 ± 0.300 a	3920 ± 0.360 a	3780 ± 0.270 a
Heart	0.351 ± 0.013 a	0.335 ± 0.003 a	0.326 ± 0.001 a

Values assigned to the letter a in the same row do not differ significantly (P > 0.05).

Table 5.
Effects of total methanolic extract of Tithonia diversifolia *and* S. occidentalis *on the relative weight of organs harvested from rats.*

3.7 Effects of extracts on biochemical and hematologic parameters

The urea level of rats treated with methanolic extract of *T. diversifolia* (0.37 ± 0.015 mmol/l) was not significantly different (ANOVA, $P > 0.05$) from the control (0.37 ± 0.026 mmol/l). However, the urea level of rats treated with methanolic extract of *S. occidentalis* (0.27 ± 0.023 mmol/l) was significantly lower (ANOVA; $P < 0.05$) than the control. Aspartate aminotransferase (ASAT) levels in the blood of rats treated with methanolic extracts of *T. diversifolia* (134.7 ± 1.76 IU) and *S. occidentalis* (131.3 ± 1.85 IU) were significantly lower (ANOVA, $P < 0.01$) than that of the control (161.5 ± 1.964 IU). Blood glucose, total cholesterol, and alanine aminotransferase (ALAT) levels observed in total extract-treated rats were not significantly different (ANOVA; $P > 0.05$) from the control (**Table 6**). In addition, hematologic analysis showed no significant change (ANOVA; $P > 0.05$) in the blood parameters of the methanolic extract-treated rats (**Table 7**).

Parameters	Control (distilled water)	*T. diversifolia* (2000 mg/kg bw)	*S. occidentalis* (2000 mg/kg bw)
Urea (mmol/l)	0.37 ± 0.026	0.37 ± 0.015	0.27 ± 0.023*
Creatinine (mmol/l)	3.66 ± 0.33	3.33 ± 0.33	3,.66 ± 0.33
Blood glucose (mmol/l)	0.97 ± 0.033	1.03 ± 0.039*	0.96 ± 0.045
Cholesterol (g/l)	0.58 ± 0.017	0.62 ± 0.014	0.56 ± 0.015
ASAT (U/I)	161.5 ± 1964	134.7 ± 1.76*	131.3 ± 1.85*
ALAT (U/I)	40.33 ± 0.88	38.33 ± 1.20	36.33 ± 0.88

*ALAT, alanine aminotransferase; ASAT, Aspartate aminotransferase. In the Table, on each line *: significant difference at P < 0.05; Absence of asterisk (*) on values indicates no significant difference p > 0.05, ANOVA, Newman–Keuls test.*

Table 6.
Effects of total methanolic extract of Tithonia diversifolia *and* S. occidentalis *on biochemical parameters of female rats.*

Parameters	Control (distilled water)	*T. diversifolia* (2000 mg/kg bw)	*S. occidentalis* (2000 mg/kg bw)
WBC (10^3/μl)	19.36 ± 1.16	20.08 ± 0.86	21.30 ± 1.35
RBC (10^6/μl)	8100 ± 0.38	7120 ± 0.27	7515 ± 0.255
HGB (g/dl)	12.93 ± 0.26	12.20 ± 0.52	12.90 ± 0.7
HCT (%)	38.90 ± 0.83	37.70 ± 0.77	38.10 ± 0.9
MCV (fL)	51.50 ± 0.81	51.53 ± 0.58	51.60 ± 0.7
MCH (pg)	17.20 ± 0.23	17.13 ± 0.12	17.05 ± 0.35
MCHC (g/dl)	31.53 ± 0.37	32.71 ± 0.32	33.00 ± 0.2
PLT (10^3/μl)	723.7 ± 17.07	710.3 ± 16.29	705.5 ± 14.5
NEUT (%)	21.45 ± 0.65	21.45 ± 0.88	21.12 ± 1.34

WBC, white blood cells; RBC, red blood cells; HGB, hemoglobin; HCT, hematocrit; MCV, mean blood volume; MCH, mean corpuscular hemoglobin; MCHC, mean corpuscular hemoglobin concentration; PLT, blood platelets; NEUT, neutrophil. The absence of an asterisk (*) on the values indicates that there is no significant difference p > 0.05, ANOVA, Newman–Keuls test.

Table 7.
Effects of methanolic extract of Tithonia diversifolia *and* S. occidentalis *on hematologic parameters of female rats.*

3.8 Effects of methanolic extracts on histological parameters

The histopathological study performed on the liver, kidney, and heart revealed no structural abnormalities (inflammation, hepatic cell necrosis, and apoptosis) in the rats treated with different doses of total methanolic extracts of *T. diversifolia* and *S. occidentalis*, compared to the sections of the organs taken from the control (**Figure 4**).

4. Discussion

In the laboratory, methanolic extracts of the leaves of *S. occidentalis* and *Tithonia diversifolia* showed contact insecticidal activity against the termite *Ancistrotermes*

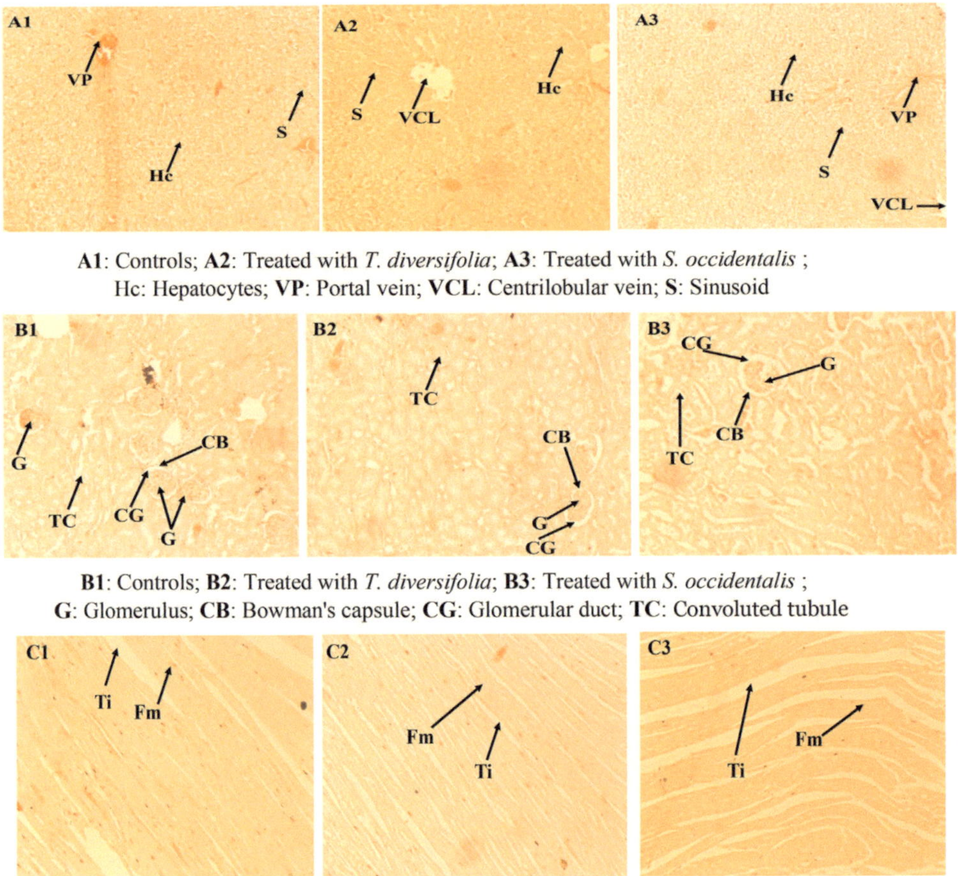

Figure 4.
Histological sections of rat liver (A), kidney (B), and heart (C). Staining: Hematoxylin and eosin, G × 100.

cavithorax. Several studies have shown contact insecticidal activities on termites by plant extracts. Extracts of *Azadirachta indica* and *Carica papaya* showed contact insecticidal activity on *Macrotermes bellicosus* [14, 18]. *C. papaya* extract also showed contact insecticidal activity on *Coptotermes formosanus* Shiraki [11]. *T. diversifolia* extract also showed insecticidal activity on *Ancistrotermes* sp. [19].

The mortality rate of *Ancistrotermes cavithorax* termites observed after treatment with total methanolic extracts increases proportionally with concentration and time. Thus, the higher the dose, the more toxic over time and the more active the products are. These results therefore suggested that methanolic extracts of *S. occidentalis* and *T. diversifolia* have their optimal action on *Ancistrotermes* termites at high dose and long term. The work of Diby et al. [19] also showed that the aqueous extract of *T. diversifolia* is toxic at high dose and long term on *Ancistrotermes* sp. These results showed that the toxicity of methanolic extract of *T. diversifolia* is significantly higher than that of *S. occidentalis* on *Ancistrotermes cavithorax* termites.

The LC50 values confirmed that the methanolic extract of *T. diversifolia* (LC50 = 29.76 mg/l) would be more active on *Ancistrotermes cavithorax* termites than the methanolic extract of *S. occidentalis* (LC50 = 84.9 mg/l).

The mortality of termites recorded during the experiment shows that the biological efficacy of the methanolic extract of *S. occidentalis* and *T. diversifolia* seems evident but is not immediate after treatment. This slow action could be due to insufficient active concentration and insufficient penetration of the product through the cuticle or oral apparatus of the insects [20, 21]. This observation is confirmed by the 50% mortality rates of *Ancistrotermes* termites obtained before 48 h with the total methanolic extract of *S. occidentalis* and *T. diversifolia*.

The extracts also seemed to act by inhalation. Indeed, inhalation with the methanolic extract of *S. occidentalis* and *T. diversifolia* significantly reduced the population of *A. cavithorax* termites tested. Contact and inhalation would thus be the two essential ways to the effectiveness of the two extracts on the *Ancistrotermes* termite.

For the barrier soil tests, the sensitivity of the *A. cavithorax* termite workers did not allow the test to continue over a long period. Only 10 days of continuous testing provided results. The results confirmed the toxicity of the methanolic extract of *T. diversifolia* and *S. occidentalis*, as well as the anti-appetitive action of these extracts on consumption. In this study, the extracts used as a barrier soil did not prevent termites from coming into contact with the food after the treated substrate chamber. Workers mortality rates in the treated substrate chambers and in the food chambers located after the treated chambers confirm that the termites came into contact with the food. Tahiri [11] and Diby et al. [19] showed that *C. papaya* and *T. diversifolia* macerates used as barrier soil did not prevent *C. formosanus* and *Ancistrotermes* sp. termites from contacting the food. In our trials, we did not obtain total mortality. Blaske and Hertel [22], Tahiri [11], and Diby et *al.* [19] also showed that compounds derived from *Jatropha cursas*, *A. indica*, *C. papaya*, and *T. diversifolia* tested as barrier soil did not cause total mortality in *Coptotermes* sp. and *Ancistrotermes* sp.

Although termites have penetrated the barrier soil (treated soil), food in the food chamber is not consumed. In nearby untreated food chambers, termite consumption is affected and is much lower than in controls. Zhu et al. [23] and Tahiri [11] showed that other botanical insecticides tested on *Coptotermes* sp. also affect termite foraging.

The acute in vivo toxicity study of the two plant extracts, 14 days after oral administration of methanolic extracts of *T. diversifolia* and *S. occidentalis* to different batches of female rats, did not record any deaths at the single dose of 2000 mg/kg body weight. The toxicity of the chemicals was determined by the LD50 value. The administered dose is 2000 mg/kg body weight, the LD50s would therefore be higher than 5000 mg/kg by the oral route. According to the OECD Globally Harmonized System of Classification and Labeling of Chemicals (2001), methanolic extracts of *T. diversifolia* and *S. occidentalis* can be classified as nonclassified category 5. The dose of 2000 mg/kg body weight would be below the maximum tolerated dose (MTD), and these extracts would therefore be nontoxic.

The results obtained in this study corroborate those obtained by other authors on the same plants. Thus, it was shown that the total methanolic extract of *S. occidentalis* leaves was not toxic in rats at a dose of 1000 mg/kg body weight 72 h after oral administration [24]. In contrast, the 70% ethanolic extract of *T. diversifolia* leaves showed a dose- and time-dependent toxic effect. This effect was reversible on the kidneys and liver, with no discernible adverse effect on the morphology of the heart, spleen, and brain [17].

Relative organ weights showed no significant difference from the control at the end of 14 days of treatment. This result suggested that the single dose of 2000 mg/kg body weight of methanolic extracts of *T. diversifolia* and *S. occidentalis* would not influence organ morphology.

Analysis of biochemical parameters showed that the creatine level of the rats did not change significantly ($P > 0.5$) compared to the control at the end of 14 days of treatment at the dose of 2000 mg/kg body weight. Only urea level decreased significantly ($P < 0.5$) in rats treated with methanolic extracts of *S. occidentalis* compared to the control. Urea and creatinine are markers of renal function that indicate glomerular filtration rate, concentrating, and diluting capacity [25]. Increased values of these markers indicate kidney dysfunction [26]. The reduction in serum urea levels in the present study has no risk to the kidney. The single dose of 2000 mg/kg bw of methanolic extracts of *T. diversifolia* and *S. occidentalis* leaves do not show any hazard on kidney.

ALAT activity also showed no significant change ($P > 0.05$) with the administration of the methanolic extracts of *T. diversifolia* and *S. occidentalis* compared with control rats. In contrast, there was a significant ($P < 0.05$) reduction in ASAT activity compared with control rats. ALAT is more specific for liver damage, whereas ASAT is predominantly present in the liver, muscle, heart, kidney, brain, and pancreas [27]. Serum levels of these enzymes increase in myopathy, rhabdomyolysis, myocardial infarction, hepatitis, or hepatic cirrhosis. The significant decrease in ASAT activity indicates that the administration of methanolic extracts of *T. diversifolia* and *S. occidentalis* at the dose studied would not cause either liver toxicity or cardiac damage. This fact is comparable to the previous diagnosis of Ejelenou et *al.* [25], according to which the administration of saponin from the leaves of *T. diversifolia* at the dose 20–100 mg/kg body weight does not cause any risk to the hepatic, renal, and cardiac tissue. In the case of this study, it could be that the extracts have a hepatoprotective action at the dose of 2000 mg/kg of body weight, which would lower the ASAT level in the liver. Indeed, several studies have highlighted the hepatoprotective role of plant flavonoids [28, 29].

Macroscopic and microscopic examinations of the liver, kidney, and heart of animals treated with different doses of methanolic extracts of *T. diversifolia* and *S. occidentalis* showed a normal structural architecture compared to the control. This confirmed that single oral administration of 2000 mg/kg body weight of *T. diversifolia* and *S. occidentalis* extracts, after 14 days, does not result in any harmful changes and morphological disorders. The absence of structural abnormalities, inflammation, hepatic cell necrosis, and apoptosis of the liver, kidney, and heart in all rats treated with methanolic extracts of *T. diversifolia* and *S. occidentalis* confirmed the results of the relative organ weights and biochemical parameters. Thus, methanolic extracts of *T. diversifolia* and *S. occidentalis* would not be toxic at a single dose of 2000 mg/kg bw in adult rats after 14 days.

5. Conclusion

This toxicity study showed that methanolic extracts of *T. diversifolia* and *S. occidentalis* were toxic by contact and inhalation on the crop pest termites, *Ancistrotermes cavithorax*. Used in barrier soil, these extracts were anti-palatable but not repellent. The research on the toxicity of these extracts on mammals, here the rat, shows that according to the OECD guideline, these two extracts were classified as nontoxic. The confirmation by the hematologic and biochemical study did not show any proven toxicity of the extracts at the single dose of 2000 mg/kg of body weight. The histopathological study carried out on the liver, the kidney, and the heart did not reveal any structural abnormality (inflammation, hepatic cellular necrosis, apoptosis) and confirmed the nontoxicity of these two extracts. These two plants could therefore be recommended to farmers as biopesticides in order to treat attacks by *Ancistrotesmes* in their crops without danger to their health.

Author details

Siapo Yao Martin and Tahiri Annick*
UFR Biosciences, Laboratory of Biology and Health, University Félix Houphouët-Boigny, Abidjan, Ivory Coast

*Address all correspondence to: tayaman2@yahoo.fr

References

[1] Watt AD, Stork NE, Mc Beath C, Lawson GL. Impact of forest management on insect abundance and damage in a lowland tropical forest in southern Cameroon. Journal of Applied Ecology. 1997;**34**:985-998

[2] Wood TG. The agricultural importance of termites in the tropic. Agricultural Zoology Reviews. 1996;**7**:117-155

[3] Akpesse AAM, Yao NGR, Coulibaly T, Diby YKS, Kouassi KP, Koua KH. Attacks and damage of termites (Insecta : Isoptera) in cocoa plantations (*Theobroma Cacao* L.) of M'Brimbo S.A.B station (South Côte D'Ivoire). International Journal of Advanced Research. 2019;**7**(1):438-445

[4] Coulibaly T, Akpesse AAM, Yapi A, Zirihi GN, Kouassi KP. Dégâts des termites dans les pépinières de manguiers du nord de la Côte d'Ivoire (Korhogo) et essai de lutte par utilisation d'extraits aqueux de plantes. Journal of Animal &Plant Sciences. 2014;**22**(3): 3454-3468

[5] Diby YKS, Tahiri YA, Akpesse AAM, Tra Bi CS, Kouassi KP. Évaluation de l'effet insecticide de l'extrait aqueux de *Tithonia diversifolia* (Hemsl.) gray (Asteracee) sur les termites en culture du riz (NERICA 1) au centre de la Côte d'Ivoire. Journal of Animal & Plant Sciences. 2015;**25**(3):3966-3976

[6] Tahiri A, Mangué J. Stratégie d'attaque des jeunes plants d'hévéa (*Hevea Brasilliensis* Muell.) par les termites et effet comparé de deux insecticides utilisés pour leur protection en basse Côte d'Ivoire. Science & Nature. 2007;**4**(1):45-55

[7] Cowie RH, Wood S. Damage to crops, forestry and rangeland by fungus-growing termites (Termitidae : Macrotermitinae) in Ethiopia. Sociobiology. 1989;**15**(2):139-153

[8] Rahimi R, Abdollahi M. A review on the mechanisms involvedin hyperglycemia induced by orgaophosphorus pesticides. Pesticide Biochemistry and Physiology. 2007;**88**:21-115

[9] Anjorwalla P, Belmain S, Sola P, Jamnadass R, Stevenson PC. Guide des plantes pesticides. Nairobi, Kenya: World Agroforestry Centre (ICRAF); 2016. p. 74

[10] Siapo YM, Tahiri A, Diby YKS, Ano EJ. Evaluation insecticidal potential of methanolic extracts of *Senna occidentalis* Link (1829) and *Tithonia diversifolia* (Hemsl) A Gray (1883) on thetermite *Ancistrotermes*. European Journal of Biotechnology and Bioscience. 2018;**6**(4):50-54

[11] Tahiri A. Toxicité du macérât de *Carica papaya* L. contre *Coptotermes formosanus* Shiraki (Isoptera : Rhinotermitidae). Afrique Science. 2012; **8**(3):93-101

[12] Mkenda P, Mwanauta R, Stevenson PC, Ndakidemi P, Mtei K, Belmain SR. Extracts from field margin weeds provide economically viable and environmentally benign pest control compared to synthetic pesticides. Public Library of Science. 2015;**10**(11):e0143530

[13] Kaushik N, Vir S. Variations in fatty acid and composition of Neem seeds collected from the Rajasthan State of India. Biochemical Society Transactions. 2000;**28**:880-882

[14] Tahiri A, Amissa AA, Adjé AF, Amusant N. Effet pesticide et screening des extraits d'*Azadirachta indica* (A.)

Juss. Sur le termite *Macrotermes bellicosus* Rambur. Bois et Forêts Tropicales. 2011; **310**(4):79-88

[15] Bounihi A. Criblage phytochimique, Étude Toxicologique et Valorisation Pharmacologique de Melissa officinalis et de Mentha rotundifolia (Lamiacées) [Thèse de Doctorat en Pharmacie]. Rabat (Maroc): Université Mohammed V; 2015. p. 199

[16] Silva Mirtes GB, Aragao TP, Vasconcelos CFB, Ferreira PA, Andrade BA, Costa IMA, et al. Acute and subacute toxicity of Cassia occidentalis L. stem and leaf in Wistar rats. Journal of Ethnopharmacology. 2011;**136**(2): 341-346

[17] Elufioyer TO, Alatise OI, Fakoya FA, Agbedahunsi JM, Houghton PJ. Toxicity studies of *Tithonia diversifolia* A. Gray (Asteraceae) in rats. Journal of Ethnopharmacology. 2009;**122**(2): 410-415

[18] Tahiri A, Amissa AA, Assi M. Toxicité et mode d'action des extraits de *Carica papaya* L. (Caricaceae) sur *Macrotermes bellicosus* Rambur (Isoptera; Macrotermitinae). Cahiers Agricultures. 2010;**19**(4):267-272

[19] Diby YKS, Tahiri A, Adja NA, Danho M, Akpesse AAM, Kouassi KP. Repellent and toxic of macerât of *Tithonia diversifolia* (Hemsl.) gray (Asteraceae) on *Ancistrotermes* sp. at the laboratory. International Journal of Multidisciplinary Research and Development. 2018;**5**(12):143-147

[20] Maistrello L, Henderson G, Laine RA. Efficacy of vetiver oil and nootkatone as soil barriers against Formosan subterranean termite (Isoptera: Rhinotermitidae). Journal of Economic Entomology. 2001;**94**: 1532-1537

[21] Zhu W, Chan EK, Li J, Hemmerich P, Tan EM. Transcription activating property of autoantigen SG2NA and modulating effect of WD-40 repeats. Experimental Cell Research. 2001;**269**(2):312-321

[22] Blaske VU, Hertel H. Repellent and toxic effects of plants extracts on subterranean termites (Isoptera: Rhinotermitidae). Journal of Economic Entomology. 2001;**94**(5):1200-1208

[23] Zhu BCR, Henderson G, Adams RP, Mao L, Yu Y, Lain RA. Repellency of vetiver oils from different biogenetic and geographical origins against *Formosan subterranean* termites (Isoptera: Rhinotermitidae). Sociobiology. 2003; **42**:623-638

[24] Irie-N'guessan AG, Kablan BJ, Kouakou-Siransy NG, Leblais V, Champy P. Evaluation de la toxicité de cinq plantes antiasthmatiques de la médecine traditionnelle ivoirienne. International Journal of Biological and Chemical Sciences. 2011;**5**(3):1316-1319

[25] Ejelonu OC, Elekofehintia OO, Adanlaw IG. *Tithonia diversifolia* saponin blood lipid interaction and its influence on immune system of normal Wistar rats. Biomedicine & Pharmacotherapy. 2017;**87**:589-595

[26] Gowda S, Desai PB, Kulkami SS, Hull VV, Math AAK, Vernekar SN. Markers of renal function tests. North American Journal of Medicine & Science. 2010;**2010**:2170-2173

[27] Goddard C, Warnes T. Raised liver enzymes in asymptomatic patients: Investigation and outcome. Digestive Disease. 1992;**10**:218-226

[28] Kebieche M. Activité biochimique des extraits flavonoïdiques de la plante Ranunculus repens L.: effet sur le diabète

expérimental et l'hépatotoxicité induite par l'Epirubicine [Thèse de Doctorat]. Algerie: Université Mentouri Constantine; 2009. p. 144

[29] Sudha A, Sumathi K, Manikandaselvi S, Prabhu NM, Srinivasan P. Anti-hepatotoxic activity of crude flavonoid fraction of Lippia nodiflora L. on ethanol induced liver injury in rats. Asian Journal of Animal Sciences. 2013;7(1):1-13

Chapter 3

Integrated Management of Termites

Lovely Bharti, Kajol Yadav and Ashok Kumar Chaubey

Abstract

Termites play an important role in the world ecosystem. Termites mostly feed on dead plant material and cellulose, generally in the form of wood, leaf litter, soil, or animal dung. These are common wood eaters, damaging the wood and wood bases of building materials. Since the supply of wood is limited, it is necessary to protect the wood and wood products from biological deterioration. For controlling termites, different management techniques are available. However, the efficiency of controlling mechanism depends on the distribution, biology, and ecology. Moreover, an integrated termite management is an environmentally safe and economically affordable mechanism of termite management. Various viable options in termite management are termite baits, the use of synthetic pesticides/insecticides/termiticides, chemicals, or botanicals. Furthermore, natural enemies include mammals, birds, insects, Araneae, microbes-bacteria, fungi, and nematodes. Here, we discuss how to utilise many integrated pest management components, so as to save crops and environment.

Keywords: termites, management, nematodes, botanicals, biology, ecology and distribution

1. Introduction

Termites, also known as 'white ants', are not true ants; they are social insects in the order Isoptera. They are essential members of the soil ecosystem (tropical and subtropical ecosystems) and are found throughout the world [1]. Termites are essential for the nutrients' cycling, decomposition of plant litter, organic matter turnover, improvement of soil fertility, providing food for other animals, and soil conditioning [2, 3]. Although they have beneficial values because of their actions, they also damage the target [4]. The undesirable change in the properties of a material caused by the vital activities of living organisms such as termites different from physical, chemical, and mechanical influences is referred to as biodeterioration. Termites can attack nearly any soft substance that contains cellulose, including wood, plants, trees, papers, homes, stones, plastics, leather, and many others [5]. It is a key component of the decomposition process and a major arthropod detritivore.

Termites are highly destructive and polyphagous pests of crop plants that cause damage to green foliage, seedlings, forests, pastures, wooden structures, fibres including domestic cellulose-based materials, and postharvest stored products [6, 7]. Termite damage is becoming more and more harmful to a wide variety of objects, particularly wooden homes, woodworking projects, buildings, agricultural plantings, trees, structural timber, manufactured wooden goods, and papers. This has resulted in enormous losses throughout the world [8]. Food security is hampered by termite damage, and smallholder farmers' livelihoods are in danger [9].

The harmful effect includes damage to human wants like crop, trees, building, and wooden materials. Therefore, the objectives of this chapter were to provide the information on termite biology, distributions, and management practice.

2. Distributions

Termites are found on all continents except Antarctica and there are several varieties of them all over the planet [10]. The tropics, subtropics, and warmer climatic areas have the densest habitats. Termites grow in warm, humid lowlands and near coastlines [11]. They predominantly occur in tropical environments, with equatorial rainforests having the maximum species richness and species diversity generally decreasing with increasing latitude. Termite species diversity and distribution differ from continent to continent as well as within a single country [12].

Termite species differ in their diversity and distribution across the world. There are only 10 species recognised in Europe and 50 in North America, while there are approximately 400 species known in South America. There are 435 different species of termites in Asia, the majority of which are found in China. Within China, Termite species only occur in areas that are mildly tropical and subtropical, south of the Yangtze River. Of the 3000 species of termites that have been identified, 1000 are found in Africa in various ecological distributions of their mounds [11].

3. Biology

Termites are tiny (4–15 mm long) and occur in a variety of colours, including white, tan, and even black [13]. The body is composed of head, thorax, abdomen, and six legs. Different-looking individuals of termites (referred to as castes) live together in the colony. The queen is the largest individual [14, 15]. Her primary function is to lay thousands of eggs per day. A king is by her side at all times. Other individuals have enormous heads with powerful jaws or heads that resemble bulbs that squirt liquid. These individuals are referred to as soldiers. However, the vast majority of termites in the colony are referred to as workers. They work long hours taking care of the queen, constructing and maintaining the nest, or gathering food and feeding the larvae and raising them.

Workers in termites, a social insect, can be either male or female. Some individuals have their wing buds grow longer. Finally, the nymphs grow into fully fledged adults, the future kings and queens. The wings are opaque grey to black, while the body colour varies from black to light brown. Swarming may occur at different times depending on the species, although it typically happens after rain. Termites are prominent in tropical ecosystems all over the world, and they are capable of creating huge earthen mounds. These mounds are air-conditioned and may contain millions of individuals [11].

3.1 Life cycle

The life cycle of termites can result in three distinct types of termites or caste types. The three types include workers, soldiers (infertile form), and a reproductive form (queen and king). The termite life cycle includes an egg, immature termite larvae or nymphs, adult nymphs, workers, soldiers, pseudergates, drones, and queens. This typical social insect system allows for adequate labour division [11]. When a termite egg among thousands of others hatches, then it is called termite larva or larvae. The larva can develop into one of the three castes, including a worker, soldier, or a secondary or supplementary reproductive termite. This occurs as a result of social, environmental, and termite pheromone cues. After that, the larva undergoes three moults until it is fully developed. The larva can also transform into worker or soldier and its life cycle ends when it dies. The larva may also develop into a reproductive alate or secondary reproductive, which allows it to continue reproducing and develop into a king or queen termite for a different colony [16]. The queen lives the longest, with an average lifespan of 25 years. The lifespan of the other types of termites ranges from 12 to 24 months.

3.2 Reproduction

Termites reproduce and replace themselves just like other organisms do to maintain their generation. As reproduction behaviour, King and Queen termites swarm in the summer in massive groups numbering in the thousands in order to find a mate [17]. The two mates performed a mild courtship dance, which started to establish their own colony. The male or king shares the labour with the queen as the queen becomes fertilised and ready to start producing termites. A queen's laying of eggs during the first year of egg production can range from 100 to thousands every day [11]. The two care for the first few generations up until there are enough young termites or workers. When eggs hatched into larvae, young termites can develop into either workers or soldiers depending on the pheromones and temperatures. The workers are the sole providers in the colony's division of labour and it depends on them to take care of all the feeding, keeping the young and developing babies in order, and go foraging. Male or female workers and soldiers are equally acceptable because they are both infertile [18]. The colony's population will continue to grow rapidly for around 5 years until the queen gives birth to her first reproductive alates, or young kings and queens. In the summer, they will become mature, get ready to swarm, and move to establish a new colony. The cycle continues to repeat itself.

4. Global status and economic importance

Currently, 3105 species of termites in 330 genera distributed over 12 families [19] and 14 subfamilies [20] are identified around the world, with about 185 of those species being considered pests [21]. Over 300 species belonging to 37 genera and 7 families have been recorded from the Indian subcontinent, of which roughly 12% (or 35 species) are known to cause harm to agricultural crops [22]. Termites, which belong to the families such as Hodotermitidae (Anacanthotermes and Hodotermes), Kalotermitidae (Neotermes), Rhinotermitidae (Coptotermes, Heterotermes, and Psammotermes), and Termitidae (Amitermes, Ancistrotermes, Cornitermes,

Macrotermes, Microcerotermes, Micromeres, Odontotermes, Procornitermes, and Syntermes), cause great damage to agricultural crops [23]. Nearly 80% of the losses in South Asia are accounted for by two species, *Odontotermes obesus* and *Microtermes obesi*. In India, termites frequently occur in red, sandy loam, lateritic, and red loam soils [20].

According to Heinrichs and Barrion [24], termites are more likely to cause damage to plants that are facing stress from biotic (insects, pests, and diseases) and abiotic (drought, nutrient deficiency, etc.) factors. Termites are becoming a serious threat to all important crops including cereals (rice, wheat, barley, maize, millet, and sorghum), pulses, oilseeds, vegetables, fruits (guava, citrus, banana, mango, papaya, grapes, etc.), sugarcane, and cotton [25–27]. In the drier regions of India, termite infestations have been documented in maize, pearl millet, pulses, sugarcane, cotton, rice, groundnut, citrus, vegetables, spices, and fruit crops [28]. The most frequently damaged cereal crops are wheat and maize [29]. Significant losses in the production of wheat and sugarcane in northern India, maize, groundnut, sunflower, and sugarcane in southern India, tea in eastern India, and cotton in western India are known. Pardeshi [30] found that crops, which were already attacked by aphids, white grubs, and hoppers, were more vulnerable to attack by termites.

Sometimes, crop yields are reduced drastically due to the losses inflicted at or near maturity which cannot be compensated [31]. Such losses threaten marginal farmers' livelihoods and impede food security [9]. Infestation due to termites was recorded up to 20–25% in rainfed and up to 10% in irrigated crops [32]. Such losses are more significant in soils with poor agronomic and physiological conditions. African Members of the Macrotermitinae, which construct the large termitaria, are responsible for over 90% of the agricultural and forestry damage; and thus, the reputation of termites as pests is associated with the presence of termitaria in crop fields [33, 34]. Globally, the estimated loss due to termite damage is about 50 billion USD (US dollars) [35], while estimates fluctuate greatly depending on the cropping practices used in various geographic areas (**Table 1**).

Regions	Annual economic losses	References
Australia	$1.5 billion	[37]
China (Mainland)	$1 billion	[15]
Fiji Islands	$1 million	[38]
France	$0.5 billion	[15]
India	$35.12 million	[21]
Indonesia	$1 billion	[39]
Japan	$0.8–1 billion	[40]
Malaysia	$10–12 million	[41]
Philippines	$100 s million	[42]
Taiwan China	$4 million	[43]
Thailand	$0.5 billion	[44]
USA	$11 billion	[35]
World	$40 billion	[45]

Table 1.
Estimates of annual economic losses caused by termites worldwide [36].

5. Management method

The management of termites is not so easy due to their dynamic living strategies which make them to cause major havoc in different crops. The majority of species are underground and have ramifications of galleries, so it might be challenging to find and approach them. A regular monitoring of fields to detect activities of termites and their damage can protect the plants from damage. Currently, chemical control is the main management strategy [46, 47]. But due to increasing concerns about their negative effects on the environment, there is a need to identify eco-friendly techniques for its management, preserving chemical control as a last option. Various commercial products available in the market for termite control are shown in **Table 2.** There are numerous Integrated Programme in Management (IPM) techniques that are available.

5.1 Botanical extracts

Using botanicals, or pesticide plants, is one of the most significant biological methods for pest management. Different plant parts, including seed, bark, leaf, fruit, root, and wood, have been used to control pests. Several solvent extracts of plants, their column chromatographic fractions, or their bioactive principles have been suggested for pest management worldwide. Botanicals are easily available as a natural resource, cost-effective alternatives for synthetic pesticides, and favourable to the environment. They serve as feeding deterrents, behavioural moderators, toxic agents, morphogenetic agents, and phagostimulants.

In order to assess the effectiveness of *Melia azedarach* extract against *M. obesi* and *O. obesus*, Qureshi et al. [49] used different concentrations of the extract in water and methanol (100, 200, and 300 ppm (parts per million)). The results showed that

Active ingredient	Termites	Tested countries
Mirex	*Mastotermes darwiniensis*, *Reticulitermes flavipes*	Australia, Canada, and the USA
Deltamethrin, sulfluramid, avermectin, and hydramethylnon	*Coptotermes formosanus*	USA
Abamectin	*C. formosanus*, *R. flavipes*, *Reticulitermes virginicus*	USA
Zinc borate	*R. flavipes*, *R. virginicus*	USA
Fipronil	*Odontotermes formosanus*, *Reticulitermes hageni*, *R. flavipes*	USA and China
Hexaflumuron (chitin synthesis inhibitor)	*Coptotermes acinaciformis*, *Coptotermes. curvignathus*, *C. formosanus*, *Coptotermes gestroi*, *Coptotermes travians*, *Heterotermes* sp., *Nasutitermes exitiosus*, *R. flavipes*, *Reticulitermes lucifugus*, *Reticulitermes speratus*, *R. virginicus*	Australia, Cayman Islands, Italy, Japan, Malaysia, Puerto Rico, and the USA
Margosan-O	*C. formosanus*	
Bioblast™	Subterranean termites	USA and Japan

Table 2.
Commercial products for termite control, available in the market [48].

Plant	Parts used	Bioactive principle	Activity
Tagetes erecta	Leaf essential oil	(Z)-β-ocimene	Termiticidal
Lepidium meyenii	Aerial parts essential oil	3-Methoxyphenylacetonitrile, benzyl thiocyanate	Feeding deterrent
Xylopia aethiopica	Fruits and seeds	Diterpenes and amides	Antifeedant
Calocedrus formosana	Leaf essential oil	T-Muurolol	100% mortality
Melauleuca spp.	Leaf essential oil	γ-Terpinene and terpinolene	Termicidal
Dipterocarpus kerrii	Resin	Sesquiterpenoids	Toxic
Adina racemosa Miq.	Bark	Benzoic acid	Toxic
Lantana camara var. *aculeata*	Leaves	Triterpenoid, 22β-acetoxylantic acid	Toxic
Thujopsis dolabrata	Wood	β-Thujaplicin and carvacrol	Toxic
Chamaecyparis pisifera	Wood	Chamaecynone and isochamaecynone	Toxic
Cryptomeria japonica	Wood	β-Eudesmol and cedrol	Toxic
Chamaecyparis obtusa	Wood	Monoterpene, sesquiterpene, and sesquiterpene alcohol	Toxic
Thujopsis dolobrata	Wood	Thujopsene	Toxic
Adina racemosa	Bark	Benzoic acid	Toxic
Cinnamomum osmophloeum	Leaf	Cinnamaldehyde	Toxic
Dipterocarpus spp.	Resin	Sesquiterpenes	Antitermitic
	Seed	Limonoids	Antifeedant
Vetiveria zizanioides	Root	Nootkatone (a sesquiterpene alcohol) and cedrene	Arrestants, feeding deterrent, repellent, and toxic
Lonchocarpus castilloi Standley	Wood	Flavonoids, castillen D and castillen E	Feeding deterrent activity
Juniperus procera	Wood	Cedrol and cedrene	Termiticidal ability
Diospyros sylvatica	Root	Quinines such as plumbagin, isodiospyrin, and microphyllone	Termiticidal ability

Table 3.
Plants, their parts, bioactive principles, and activity against different termites [48].

O. obesus was resistant to the solvents water and methanol. Aqueous root extracts of *Crotalaria burhia* and powdered leaf dusts of *Anacardium occidentale* were tested in the laboratory for their efficacy to repel *O. obesus*. According to the research conducted by Bajya et al. [50], *C. burhia* at 10 and 20% was found to repel 66.7 and 70% of *O. obesus*, whereas *A. occidentale* leaf dust at 5 and 7.5% was found to repel 56.7 and 60% of termites. Garlic, neem, and tobacco solutions at three different concentrations (20, 10, and 5%) were tested in a laboratory condition against *Heterotermes indicola*. Garlic and tobacco concentrations resulted in high and low mortality [51]. The most common plants used for various termite control methods are shown in **Table 3.**

5.2 Mechanical and physical control

Finding termite mounds in or near crop fields and manually removing them through digging or flooding is an effective way to disrupt termites' everyday activities. On the other side, dequeening greatly aids in disrupting the life cycle but it is a labour-intensive, time-consuming process that needs trained workers [52]. After the queen has been removed, the colony must be quickly destroyed by lighting a fire. A further method used in Indian communities to suffocate termite mounds is burning crop debris on top of them. According to Rath and Bhanja [53], termite winged individuals can be gathered by poultry farmers during swarming for use as chicken feed.

5.3 Chemical control

Synthetic pesticides are artificially made compounds that have a long half-life in the environment [54]. They are effective for the protection of termites, even though they cause serious human health and environmental problems [55–57]. For instance, when methyl bromide is sprayed, it quickly enters inside wooden structures and kills arboreal species or dry wood termites [21]. Although methyl bromide effectively eliminates termites, it is thought to be a contributor to the ozone layer's deterioration [21, 58]. Additional industrial chemicals that have been used for controlling termites in stored wood, wooden components of buildings, and crops are chlorpyrifos, fipronil, bifenthrin, imidacloprid, endosulfan, lindane, aldrin, chlordane, dieldrin, endrin, and heptachlor [58–61]. Synthetic pesticides have caused a reduction in insect population in the United States of America [62] and bird populations in India [63].

In India, the most frequently used insecticide for termite management is chlorpyrifos. Farmers, all over the country, use chlorpyrifos and other conventional chemicals with irrigation water, regardless of the type of crop planted. However, scientists from many different regions of the world evaluated a wide range of novel insecticides, many of which had comparatively fewer negative effects. In northeast India, Bhagawati et al. [64] studied the field efficacy of some newer insecticides (thiamethoxam25 WG, imidacloprid 17.8 SL, acephate 50% + imidacloprid 1.8% SP, thiamethoxam35 FS, imidacloprid 600 FS, and clothianidin 50 WDG) against *O. obesus* in preserved setts and found clothianidin 50WDG to show the greatest efficacy. In Thailand, Charoenkrun [65] also found clothianidin as the most effective insecticide against subterranean termites even at a very low concentration. They reported that damage was decreased by up to 5% with an increase in dosage of 0.15%. According to Smith et al. [66], acetamiprid and bifenthrin were more toxic when used together than when used alone. Another effective method is baiting, in which the contaminated individual uses trophallaxis to spread the chemical throughout its colony. Hexaflumuron and flufenoxuron bait matrixes were found to be highly effective in eliminating subterranean termites [67, 68].

5.4 Biological control

Various entomopathogenic fungi, such as *Beauveria*, *Metarhizium*, and *Aspergillus*; bacteria, such as *Bacillus*, *Serratia*, and *Pseudomonas*; and nematodes, such as *Heterorhabditis*, *Steinernema*, and *Neosteinernema*, had been reported as biocontrol

Biocontrol agent(s)	Target species	Mortality (%)	References
Entomopathogenic fungi			
Metarhizium anisopliae	*O. obesus*, *Odontotermes guptai*	16–100	[71, 72]
	Macrotermes spp.	60v100	[73]
	Psammotermes hypostoma	22.5	[74]
	Microcerotermes diversus	20–100	[75]
	O. formosanus	100	[76]
	Termites in tea	85.97	[77]
Beauveria bassiana	*Macrotermes* spp.	25–95	[73]
	Cornitermes cumulans	100	[78]
	P. hypostoma	15	[74]
Aspergillus niger	*P. hypostoma*	13.3	[74]
EPNs			
Symbiobacterium thermophilum	*O. obesus*	44–78	[26]
Steinernema siamkayai	*O. obesus*, *R. flavipes*	80–100	[79]
Steinernema pakistanense			
S. pakistanense	*Macrotermes* spp.	50–96	[80]
Steinernema carpocapsae	*M. diversus*	37.5–96.2	[81]
Steinernema riobrave	*Heterotermes aureus*, *R. flavipes*, *Gnathamitermes perplexus*	≥80	[82]
Heterorhabditis indica	*O. obesus*, *R. flavipes*	20–100	[79]
Heterorhabditis baujardi, *H. indica*	*P. hypostoma*, *Aspergillus ochraceus*	32.6–95.8	[83]
Hemidesmus indicus	*Odontotermes horni*, *Odontotermes redemanni*	100	[84]
Heterohabditis bacteriophora	*M. diversus*	22.5–80	[81]
Entomopathogenic bacteria			
Bacillus thuringiensis	*M. obesi*, *Microcerotermes beesoni*	>80	[85]
B. thuringiensis sub sp. *israelensis*			
Pseudomonas fluorescens	*O. obesus*		[86]
Serratia marcescens	*C. formosanus*	100	[87]

Table 4.
Important biocontrol agents for termite management [70].

agents [57, 69] (**Table 4**). Myles [88] listed 2 viruses, 5 bacteria, 17 fungi, 5 nematodes, and 4 mites as natural enemies. These biocontrol agents offer an environment-friendly and cost-effective control [89]. Under natural conditions, diseased colonies are rarely encountered, as they maintain a very strict sanitary regimen, with absolute cleanliness, by removing the diseased and dead individuals from the colony. They may also produce selective antibiotics and ensure that only Termitomyces grow in their fungal gardens. Therefore, the management using biocontrol agents requires external applications of these natural enemies.

5.4.1 Entomopathogenic fungi

A plethora of information is available regarding entomopathogenic fungi's ability to infect termites [90]. Entomopathogenic fungi protect the crop through two different ways. The first is direct killing brought on by contact with fungal conidia, and the second is avoidance brought on by spores's repulsive effects [91]. Rath [90] also noted that termites have the potential to transmit the fungus inside the colony through grooming and other social interactions, such as trophallaxis, demonstrating the efficacy of colony management through the treatment of distant feeding sites. *Metarhizium anisopliae* (Metsch.) and *Beauveria bassiana* (Bals.) are two fungal species that have received the majority of attention in biocontrol studies until now [92]. In Karnataka, Balachander et al. [71] examined how workers and soldiers of *O. obesus* were affected by dried fungal suspensions of *M. anisopliae* at 108 conidia/ml combined with attractants (100 g mixture of sugarcane bagasse, cardboard, and sawdust). The mortality rates among workers and soldiers as a result of the horizontal spread of *M. anisopliae* infection were 50–98 and 16–78%, respectively.

In Khuzestan Province, Cheraghi et al. [75] evaluated the effectiveness of a bait matrix treated with *M. anisopliae*, Strain Saravan (DEMI 001), against *M. diversus*. The results showed that concentrations of 3.7×10^7 and 3.5×10^8 conidia/ml had the highest mortality rates. In Kenya, Maniania et al. [93] used granular formulation of *M. anisopliae* against maize termites under field conditions and observed significant reduction in lodging and an improvement in crop production. In China, Hussain et al. [94] examined conidial suspensions of *M. anisopliae* strain ARSEF 6911 (1×10^{10}, 1×10^8, 1×10^6, and 1×10^4 conidia/ml) mixed with diesel oil against *M. obesi* and *O. obesus* and discovered a significant reduction in infestation. According to Hoque et al. [77], the fungal formulation Bioterminator (*M. anisopliae*) was proven to be highly effective in tea in Bangladesh. In Kerala, Swaran and Varma [72] examined *M. anisopliae* var. *major* at 0.8×10^7, 1.2×10^7, 1.6×10^7, and 2×10^7 conidia/g soil under laboratory conditions and outcomes revealed 100% mortality of *Odontotermes guptai* Roonwal and Bose in 192, 156, 132 and 120 hours, respectively. In China, Dong et al. [76] studied the effectiveness *of M. anisopliae* var. *dcjhyium* against *O. formosanus* under laboratory conditions and found 100% mortality.

In Rajasthan, Intodia et al. [95] used histological procedure to assess the likelihood that *B. bassiana* (Balsamo) Vuillemin (4.5×10^8 conidia/ml) would infect the midgut of *O. obesus* workers. Transverse body sections showed that the fungus entered the gut musculature and caused termite mortality. In Ethiopia, the use of *M. anisopilae* and *B. bassiana*, two entomopathogenic fungi, at higher concentrations for 7 days was found to be both successful and beneficial to the environment, with >95% mortality of *Macrotermes* spp. The mortality of *Macrotermes* spp. varied from 60 to 100 and 25 to 95% due to *M. anisopliae* and *B. bassiana*, respectively [73]. Fernandes and Alves [78] observed 100% mortality of *Cornitermes cumulans* Kollar within 10 days of application of dust formulation of *M. anisopliae* and *B. bassiana* under field conditions. Singha et al. [96] with two strains each of *M. anisopliae* Sorokin and *B. bassiana* Vuillemin observed their potential as biological control agents against the tea termite, *M. obesi*; and *M. anisopliae* strains were found to be more virulent with lower concentration than *B. bassiana* strains. Soliman et al. [74] tested the efficacy of entomogenous fungi against of *P. hypostoma* termites in Egypt and found *M. ansiopliae* (mortality: 22.5%) to be more effective than *B. bassiana* (mortality: 15.0%) and *A. niger* (mortality: 13.3%).

5.4.2 Entomopathogenic nematodes

According to Chouvenc et al. [97], entomopathogenic nematodes (EPNs) have received less attention than entomopathogenic fungi in research on the biological control of subterranean termites worldwide. The first report on the use of nematodes was reported by Tamashiro [98] in Hawaii, and many other researchers followed this method of action. EPNs from the families such as Steinernematidae and Heterorhabditidae are effective alternatives for controlling subterranean termites because they are able to search for, infect, and kill the host insect [99]. The EPNs attack many termite body parts, including their nervous tissue, muscle tissue, fat body, salivary gland, and sternal gland [100]. According to Qasim et al. [100], 30 isolates and about 34 EPN species have been discovered to have a parasitic relationship with termites worldwide. These organisms are used to control termites. They are very effective against termites under laboratory conditions, but their effectiveness in the field has not been confirmed.

Rathour et al. [26] assessed a native Indian EPN *S. thermophilum*-based product called 'PusaNemagel' for the management of *O. obesus* in wheat and pearl millet crops in India. The findings showed that termite incidence in wheat and pearl millet may be decreased by 48–78 and 44%, respectively. Poinar [101] observed mortality among *Coptotermes* and *Nasutitermes* workers when challenged with *Neoaplectana carpocapsae* (Steiner) (=*Steinernema feltiae*) in laboratory experiments. Similar results were achieved by Georgis et al. [102] using *N. carpocapsae* and *Heterorhabditis* nematodes in *Zootermopsis* and *Reticulitermes*. El-Bassiouny and El-Rahman [83] examined the parasitisation capability of *H. baujardi* and *H. indica* (Poinar) against Egyptian termites, *P. hypostoma* and *A. ochraceus*, under laboratory conditions. They concluded that *P. hypostoma* was more vulnerable to *H. baujardi* than to *H. indica*. In Sri Lanka, Danthanarayana and Vitarana [103] successfully managed damp wood termite, *Glyptotermes dilatatus* (Bugnion and Popoff), using *Heterorhabditis* sp. in tea plantations. In Tamil Nadu, Razia and Sivaramakrishnan [79] tested the pathogenicity of three species of EPNs, *Steinernema siamkayai* Stock, *S. pakistanense*, and *H. indica*, against subterranean termites. They also found that *S. pakistanense* at 250 IJs (infective juveniles)/ml showed 100% mortality of both *Reticulitermes flavipes* (Kollar) and *Odontotermes horni* within 24 hr. than *S. siamkayai* and *H. indica*.

Wang et al. [99] examined the effectiveness of various EPNs against *R. flavipes* and *C. formosanus*. The order of effectiveness was *H. indica* > *H. bacteriophora* > *S. carpocapsae* > *S. riobrave*. In Arizona (USA), Yu et al. [82] examined the biocontrol potential of *S. riobrave*, *S. carpocapsae*, *S. feltiae*, and *H. bacteriophora* against *Heterotermes aureus* (Snyder), *R. flavipes*, and *Gnathamitermes perplexus* (Banks) in laboratory sand experiments. The results proved that all EPNs were capable of infecting and killing different kinds of termites; however, *S. riobrave* displayed the maximum mortality (80%) of both species. Shahina and Tabassum [80] documented a substantial amount of mortality in *Macrotermes* species due to *S. pakistanense* using filter paper and sand assay under laboratory conditions in Pakistan. According to Dolinski and Lacey [104], *Neotermes* spp. that infect palm trees in the South Pacific Islands can be controlled using *Heterorhabditis* sp. Zaidawi et al. [81] tested three EPN species against *M. diversus* in the field. They discovered that *H. bacteriophora*, *S. carpocapsae*, and *H. bacteriophora* (IRQ.1 strain) caused mortality rates of termite workers 22.5–80 ± 8.3, 37.5–96.2 ± 8.9, and 28.7–67.5 ± 6.8%, respectively. In Sri Lanka, *H. indicus* at 1000 IJs/ml showed 100% mortality of *O. horni* and *O. redemanni* in laboratory conditions [84].

5.4.3 Entomopathogenic bacteria

Entomopathogenic bacteria were the first biological agents that were examined for the control of termites [105]. Among other bacterial entomopathogens, *Bacillus* spp. have been used as important biocontrol agents for termite management. In Assam, Singha et al. [85] examined the pathogenicity of *B. thuringiensis* and *B. thuringiensis* subsp. *israelensis* against two species of tea termites, *M. obesi* and *M. beesoni*. They found more than 80% mortality in both the termite species. However, it was found that *B. thuringiensis* subsp. *israelensis* was more virulent than *B. thuringiensis*. Omoya and Kelly [106] isolated two entomopathogenic bacteria (*B. subtilis* and *S. marcescens*) from macerated termite cadavers and analysed their pathogenicity against *Macrotermes bellicosus*. According to their findings, termites could be degenerated by *B. subtilis* and *S. marcescens* at concentrations of 108 and 107 cfu (colony-forming units)/ml, respectively. *P. fluorescens* (a rhizobacterium that produces hydrogen cyanide) was also discovered to be highly effective against *O. obesus* under laboratory conditions in New Delhi [86]. The most promising bioagent among 15 bacterial species, *S. marcescens*, gave 100% mortality of termites, *C. formosanus*, under laboratory conditions in the United States of America [87].

5.5 Integrated termite management and traditional practices

5.5.1 IPM practices

Integrated termite management uses every possible combination of termite control methods. [107]. Ramakrishnan et al. [108] found that the susceptibility of *R. flavipes* to *M. anisopliae* was increased in the presence of imidacloprid. Yii et al. [109] examined on the compatibility and synergy in efficacy of the termiticide fipronil with *M. anisopliae*, either alone or in combination, against the subterranean termite *C. curvignathus*. The results of a compatibility test revealed that sublethal doses of fipronil were relatively less harmful to fungal growth. The most powerful synergistic effects were seen in the fungus-insecticide bait formulation, which boosted termite mortality and decreased the lethal period at a sublethal dose. The termites' insecticidal stress due to the sublethal fipronil in the bait formulation may weaken their defence mechanisms, which makes it easier for fungus to infect them. The observed synergism treatments indicated a potential for the development of an integrated fungus-insecticide control approach. In another study, the combination of *M. anisopliae* with diesel oil considerably decreased insect damage by achieving higher germination (>55%) and lower bud damage (5.50%) at both sites, in both seasons. According to the findings, diesel oil and *M. anisopliae* application may be a beneficial treatment strategy for the control of sugarcane termites [94].

5.5.2 Traditional practices

Generally, most of the traditional methods of pest management are frequently used in developed countries, but these practices still play a significant role in pest control programmes in developing countries. The use of plant resources in the management of insect pests, and especially in termite control, has been an old tactic in the African culture [110]. Traditionally, in the Gushegu District of Northern Ghana, farmers used various methods to decrease termite attack, including the cultivation of elephant grass, the burial of animal and plant corpses, the use of wood ash, and the application of a mixture of salt and shea butter leftovers [111]. The effectiveness of conventional methods was evaluated by Nwilene et al. [112] using two plant extracts (neem seed oil and neem powder), which revealed that the seed oil was superior to the powder for termite control.

6. Conclusions

Termites are highly polyphagous and most destructive pests causing damage to a wide range of agricultural, horticultural, and forest plants as well as storage materials. They have the potential to cause yield losses of up to 100% when damage occurs during the crop's early growth phases. Many farmers only use chemical treatment as a method of pest management. However, the outcomes are not permanent and are also more expensive for farmers. Termites cannot be entirely eliminated or prevented in harvested areas, nor is this a good idea. There are many indigenous traditional practices, but they are generally unsuccessful and only applicable locally. So, they cannot provide better outcomes if we use them as a sole practice. These practices should be included in a well-planned IPM system to reduce the pesticide load on the ecosystem. Scientists have explored methods that are farmer-friendly, such as cultural, physical, mechanical, and biological control (entomopathogenic fungus, bacteria, nematodes, botanical extracts, etc.) and that have a wide range of possible applications for termite management. One can achieve better results by including entomopathogenic agents as part of an IPM strategy. Biocontrol agents never received serious consideration for their utilisation especially for field application. Therefore, much more study is required to examine the potential of bioagents for the management of termites infesting essential crops for agriculture. Entomopathogens can be used to effectively and selectively control insects as part of an integrated strategy. In the near future, a sustainable termite control solution will be provided through the synergistic integration of many practices.

Conflict of interest

The authors declare no conflict of interest.

Author details

Lovely Bharti*, Kajol Yadav and Ashok Kumar Chaubey
Nematology Laboratory, Department of Zoology, Chaudhary Charan Singh University, Meerut, UP, India

*Address all correspondence to: lovely13589@gmail.com

References

[1] Abdel G, Skai E. Termite damage to buildings: Nature of attacks and preventive construction methods. American Journal of Engineering and Applied Sciences. 2011;**4**:187-200

[2] Jouquet P, Traoré S, Choosai C, Hartmann C, Bignell D. Influence of termites on ecosystem functioning. Ecosystem services provided by termites. European Journal Soil Biology. 2011;**47**(4):215-222. DOI: 10.1016/j.ejsobi.2011.05.005

[3] Brauman A, Majeed MZ, Buatois B, Robert A, Pablo AL, Miambi E. Nitrous oxide (N_2O) emissions by termites: does the feeding guild matter? PLoS One. 2015;**10**:e0144340

[4] Debelo DG, Degaga EG. Preliminary study on termite damage on rural houses in the Central Rift Valley of Ethiopia. African Journal of Agricultural Research. 2014;**9**:2901-2910

[5] Verkerk R. Building out Termites: an Australian Manual for Environmentally Responsible Control. Sydney: Pluto Press Australia Limited; 1990

[6] Pearce MJ. Termites: Biology and Pest Management. 1st ed. Chatham: CAB International; 1997. p. 172

[7] Upadhyay RK. Effects of plant latex based anti-termite formulations on Indian white termite Odontotermes obesus (Isoptera: Odontotermitidae) in sub-tropical high infestation areas. Open Journal of Animal Sciences. 2013;**3**(4):281-294

[8] Ahmed S, Qasim M. Foraging and chemical control of subterranean termites in farm buildings in Faisalabad, Pakistan. Pakistan Journal of Life and Social Sciences. 2011;**9**:58-62

[9] Demissie G, Mendesil E, Diro D, Tefera T. Effect of crop diversification and mulching on termite damage to maize in western Ethiopia. Crop Protection. 2019;**124**:104723

[10] Eggleton P. Global patterns of termite diversity. In: Abe T, Bignell DE, Higashi M, editors. Termites: Evolution, Sociality, Symbioses, Ecology. The Netherlands: Kluwer Academic Publishers; 2000. pp. 25-51

[11] Eiseman C, Noah C, John C. Tracks & Sign of Insects & Other Invertebrates: Guide to North American Species. Mechanicsburg, PA: Stackpole, Print; 2010

[12] Engel MS, Pan AD, Jacobs BF. A termite from the Late Oligocene of northern Ethiopia. Acta Palaeontologica Polonica. 2012;**58**(2):331-335

[13] Nkunika POY, Shiday BA, Sileshi GW, French JRJ, Nyeko P, Jain S. Termite taxonomy and distribution with particular reference to climate change in Africa. In: 41st Annual Meeting of the International Research Group on Wood Protection, Biarritz, 9-13 May 2010; France, IRG Secretariat. 2010

[14] Krishna K. Taxonomy, phylogeny and distribution of termite. In: Krishna K, Weesner FM, editors. Biology of Termites. Vol. 1. New York/London: Academic Press, Elsevier; 1969. p. 127-152

[15] Lenz M, Sunden-Bylehn A, Thorne BL, Lewis VR, Haverty M. Finding alternatives to persistent organic pollutants (POPs) for termite

management. UNEP/FAO/Global IPM Facility Expert Group on Termite Biology and Management, United Nations Environment Programme/Food and Agriculture Organization of the United Nations/Global Integrated Pest Management Facility. United Nations Environment Programme. 2003

[16] Evans Arthur V. National Wildlife Federation Field Guide to Insects and Spiders & Related Species of North America. New York: Sterling Publication; Print; 2007

[17] Matsuura K, Fujimoto M, Goka K. Sexual and asexual colony foundation and the mechanism of facultative parthenogenesis in the termite Reticulitermes speratus (Isoptera, Rhinotermitidae). Insectes Sociaux. 2004;**51**(4):325-332

[18] Wang C, Zhou X, Li S, Schwinghammer M, Scharf ME, Buczkowski G, et al. Survey and identification of termites (Isoptera: Rhinotermitidae) in Indiana. Annals of the Entomological Society of America. 2009;**102**(6):1029-1036

[19] Engel M, David G, Kumar K. Termites (Isoptera): Their phylogeny, classification, and rise to ecological dominance. American Museum Novitates. 2009;**3650**:1-27

[20] Varshney AK. Household pests: termites (The white ants). Applied Entomology- Insecticidal methods of pest control, Zoology Department, Hindu College, Delhi; 2007. pp. 1-13

[21] Verma M, Sharma S, Prasad R. Biological alternatives for termite control: a review. International Biodeterioration and Biodegradation. 2009;**63**:959-972

[22] Murthy S, Rajeshwari K, Ramya R, Jalali T, Verghese A. Genetic diversity among Indian termites based on mitochondrial12S rRNA gene. European Journal of Zoological Research. 2015;**4**:1-6

[23] Anonymous. Global IPM Facility Expert group on termite biology and management. Finding alternatives to persistent organic pollutants (POPs) for termite management. 2000. Available from: www.chem.unep.ch/pops/termites/termite_ch4.htm

[24] Heinrichs EA, Barrion AT. Rice-Feeding Insects and Selected Natural Enemies in West Africa: Biology, Ecology, Identification. Los Banos, Philippines: International Rice Research Institute; 2004. p. 242

[25] Lin Y, Fang D, Wang L. Termites and microbial biological control strategies. Muhammad Qasim. South Asia Journal of Multidisciplinary Studies. 2015;**1**:33-62

[26] Rathour KS, Ganguly S, Das T, Singh P, Kumar A, Somvanshi VS. Biological management of subterranean termites (Odontotermes obesus) infesting wheat and pearl millet crops by entomopathogenic nematodes. Indian Journal of Nematology. 2014;**44**:97-100

[27] Sen-Sarma PK. Ecology and biogeography of the termites of India. Forest Entomology. Thakur ML, editor. Dehradun: Sai Publication; 2000. 422 p

[28] Roonwal ML. Termite Life and Termite Control in Tropical South Asia. Jodhpur, India: Scientific Publishers, United Book Traders; 1979

[29] Shrivastava SK, Verma RK, Singh B. Integrated pest management in wheat. In: Wheat: Recent Trends on Production Strategies of Wheat in India. Jabalpur and ICAR-IIWBR: Jawaharlal Nehru Krishi Vishwa Vidyalaya (JNKVV); 2014. pp. 197-209

[30] Pardeshi MV. Insect pests, their control and management in agroecosystem of Gujarat: With special emphasis on termites [PhD thesis]. Concise summary; 2007. p. 14

[31] Verma AN, Kashyap RK. Termites their damage and control in field crops. Memoir No. 8, Entomological Society of India, New Delhi; 1980. p. 53

[32] Sharma AK, Sahaan MS, Babu KS. Wheat crop health. Newsletter. 2009;**14**:23-27

[33] Abdulahi A, Abraham T, Mohammed D. Importance and management of termites in Ethiopia. Pest Manage. Ethiopian Journal of Health Sciences. 2010;**14**:1-20

[34] Sileshi G, Nyeko P, Nkunika P, Sekematte B, Akinnifesi F, Ajayi O. Integrating ethno-ecological and scientific knowledge of termites for sustainable termite management and human welfare in Africa. Ecology and Society. 2009;**14**(1):48

[35] Subekti N, Yoshimurab T, Rokhmanc F, Masturd Z. Potential for subterranean termite attack against five bamboo species in correlation with chemical components. Procedia Environmental Sciences. 2015;**28**:783-788

[36] Ahmad F, Fouad H, Liang SY, Hu Y, Mo JC. Termites and Chinese agricultural system: Applications and advances in integrated termite management and chemical control. Insect Science. 2021;**28**(1):2-20. DOI: 10.1111/1744-7917.12726

[37] Staunton I. Termite Cost: The Real Story About Termite Damage in Australia. 2012. Available from: https://termikill.com.au/termite-cost/

[38] Chand RR, Jokhan AD, Charan H, Raj K, Singh P. Threats posed by Asian subterranean termites in the Fiji Islands and their potential controls: a review. New Zealand Plant Protection. 2018;**71**:129-139

[39] Hadi Y, Massijaya M, Arinana A. Subterranean termite resistance of polystyrene-treated wood from three tropical wood species. Insects. 2016;7:37

[40] Tsunoda K, Yoshumura T. Current termite management in Japan. In: Proceedings of 1st Pacific-Rim Termite Research Group. 8-9 March 2004, Penang, Malaysia; 2004. pp. 1-5

[41] Yeoh BH, Lee CY. Tunneling activity, wood consumption and survivorship of Coptotermes gestroi, Coptotermes curvignathus and Coptotermes kalshoveni (Isoptera: Rhinotermitidae) in the laboratory. Sociobiology. 2007;**50**:1087-1096

[42] Acda MN. Geographical distribution of subterranean termites (Isoptera) in economically important regions of Luzon, Philippines. The Philippine Agricultural Scientist. 2013;**96**:205-209

[43] Li J, Hu Y, Deng TF, Guo JQ, Gong YG, Mo JC. Laboratory and field evaluation of gravel sand as a physical barrier against Coptotermes formosanus and Reticulitermes flaviceps (Isoptera: Rhinotermitidae). Sociobiology. 2011;**57**:71-78

[44] Vongkaluang C. Current termite management in Thailand. In: Proceedings of 1st Pacific-Rim Termite Research Group; 8-9 March 2004, Penang, Malaysia. 2004. pp. 43-46

[45] Rust MK, Su NY. Managing social insects of urban importance. Annual Review of Entomology. 2012;**57**:355-375

[46] Peterson C, Wagner TL, Mulrooney JE, Shelton T. Subterranean Termites-Their Prevention and Control in Buildings. USDA Home and Garden Bulletin; 2006. p. 64

[47] Potter MF. Termites. In: Mallis A et al., editors. Handbook of Pest Control: The Behavior, Life History, and Control of Household Pests. 10th ed. Cleveland: GIE Media Inc.; 2011. pp. 1600

[48] Sahayaraj K. Bio-intensive integrated management of termites. In: Khan M, Ahmad W, editors. Termites and Sustainable Management. Sustainability in Plant and Crop Protection. Cham: Springer; 2018. DOI: 10.1007/978-3-319-68726-1_7

[49] Qureshi NA, Ashraf A, Afzal M, Ullah N, Iqbal A, Haleem S. Toxic potential of Melia azedarach leaves extract against Odontotermes obesus and Microtermes obesi. International Journal of Biosciences. 2015;**6**:120-127

[50] Bajya D, Manoharan T, Sridharan S, Kuttalam S. Repellent efficacy of Crotalaria burhia and Anacardium occidentale against Odontotermes obesus (Isoptera: Termitidae) under laboratory conditions. Indian Journal of Agricultural Science. 2015;**85**:1234-1236

[51] Ahmed N, Huma Z, Rehman SU, Ullah M, Ahmed S. Effect of different plant extracts on termite species (Heterotermis indicola). Journal of Bioresource Management. 2016;**3**:9-16

[52] Khan MA, Ahmad W, Paul B, Paul S, Khan Z, Aggarwal C. Entomopathogenic nematodes for the management of subterranean termites. In: Hakeem KR et al., editors. Plant, Soil and Microbes. Switzerland: Springer International Publishing; 2016. pp. 317-352

[53] Rath AC, Bhanja PK. Farmers' indigenous traditional knowledge for pest management. In: Proceedings. National symposium on sustainable pest management for safer environment. OUAT, Bhubaneshwar, Orissa; 2007. pp. 72-74

[54] Chaudhary S, Kanwar RK, Sehgal A, Cahill DM. Progress on Azadirachta indicabased biopesticides in replacing synthetic toxic pesticides. Journal Frontiers in Plant Science. 2017;**8**(610):1-13

[55] Forschler BT, Townsend AL. Mortality of eastern subterranean termites (Isoptera : Rhinotermitidae) exposed to four soils treated with termiticides. Journal of Economic Entomology. 1996;**89**(3):678-681

[56] Kamble ST, Ogg CL, Gold RE, Vance AD. Exposure of applicators and residents to Chlordane and Heptachlor when used for subterranean termite control. Archives of Biological Sciences. 1992;**22**:253-259

[57] Muralidhara S, Prasad B, Singh R. Colonization and antagonistic activity of entomopathogenic Aspergillus sp. against tea termite (Microcerotermes beesoni Snyder). Current Science. 2013;**105**:1216-1619

[58] Himmi SK, Tarmadi D, Ismayati M, Yusuf S. Bioefficacy performance of neem-based formulation on wood protection and soil barrier against subterranean termite, Coptotermes gestroi Wasmann (Isoptera : Rhinotermitidae). Procedia Environmental Sciences. 2013;**17**:135-141

[59] Parman V, Vargo EL. Colony-level effects of imidacloprid in subterranean termites (Isoptera : Rhinotermitidae). Journal of Economic Entomology. 2010;**103**(3):791-798

[60] Rakshiya YS, Verma MK, Sindhu SS. Efficacy of antagonistic soil bacteria in management of subterranean termites (Isoptera). Research in Environment and Life Science. 2016;**9**(8):949-955

[61] Su NY. Overview of the global distribution and control of the Formosan subterranean termite. Sociobiology. 2003;**41**(1(A)):7-16

[62] Johnson RM, Ellis MJ, Mullin CA, Frazier M. Review article pesticides and honey bee toxicity - USA. Apidologie. 2010;**41**(3):312-331

[63] Mitra A, Chatterjee C, Mandal FB. Synthetic chemical pesticides and their effects on birds. Research Journal of Environmental Toxicology. 2011;**5**(2):81-96

[64] Bhagawati S, Bhattacharyya B, Mishra H, Gogoi D. Chemical management of termites (Odontotermes obesus) in preserved setts of sugarcane (Saccharum officinarum). Journal of Entomology and Zoology Studies. 2017;**5**(1):856-859

[65] Charoenkrun KA. Field trial in Thailand to evaluate the efficacy of a new non-repellent termiticide (clothianidin: neonicotinoid) in preventing structural damages caused by subterranean termite. In: Proceedings. 10th Pacific termite research group conference. BT Forschler (ed.), Kuala Lumpur, Malaysia. 2014. pp. 7-11

[66] Smith JA, Pereira RM, Koehler PG. Relative repellency and lethality of the neonicotinoids thiamethoxam and acetamiprid and an acetamiprid/bifenthrin combination to Reticulitermes flavipes termites. Journal of Economic Entomology. 2008;**101**:1881-1887

[67] Ahmed Shiday BM, French JRJ. The efficacy of flufenoxuron as potential bait toxicant against termites. African Journal of Agricultural Research. 2013;**8**:3688-3694

[68] Su NY, Thoms EM, Ban PM, Scheffrahn RH. Monitoring/baiting station to detect and eliminate foraging populations of subterranean termites (Isoptera: Rhinotermitidae) near structures. Journal of Economic Entomology. 1995;**88**:932-936

[69] Kanzaki N, Giblin-Davis RM, Herre EA, Scheffrahn RH, Center BJ. Pseudaphelenchus vindai n. sp. (Tylenchomorpha: Aphelenchoididae) associated with termites (Termitidae) in Barro Colorado Island, Panama. Nematology. 2010;**12**:905-914

[70] Rana A, Chandel RS, Verma KS, Joshi MJ. Termites in important crops and their management. Indian Journal of Entomology. 2021;**83**:e20115

[71] Balachander M, Remadevi OK, Sasidharan TO. Dissemination of Metarhizium anisopliae infection among the population of Odontotermes obesus (Isoptera: Termitidae) by augmenting the fungal conidia with attractants. Journal of Asia-Pacific Entomology. 2013;**16**:199-208

[72] Swaran PR, Varma RV. Laboratory evaluation of the entomopathogenic fungus, Metarhizium anisopliae var. major against the subterranean termite, Odontotermes guptai Roonwal and Bose. Journal of Biological Control. 2003;**17**(2):147-151

[73] Sileshi A, Sori W, Dawd M. Laboratory Evaluation of Entomopathogenic Fungi Metarhizium anisophilae and Beauveria bassiana Against Termite, Macrotermes (Isoptera: Termitidae). Asian Journal of Plant Sciences. 2013;**12**:1-10

[74] Soliman MM, Allah RK, Hammad M, Ebnalwaled AA. Biological control of subterranean termites (Psammotermes hypostoma) by entomopathogenic fungi. Scientific Journal of Agricultural Sciences. 2019;**1**(1):21-29

[75] Cheraghi A, Habibpour B, Mossadegh MS. Application of bait treated with the entomopathogenic fungus Metarhizium anisopliae (Metsch.) Sorokin for the control of Microcerotermes diversus. Psyche: A Journal of Entomology. 2013;**2013**:1-5

[76] Dong C, Zhang J, Chen W, Huang H, Hu Y. Characterization of a newly discovered China variety of Metarhizium anisopliae (M. anisopliae var. dcjhyium) for virulence to termites, isoenzyme, and phylogenic analysis. Microbiological Research. 2007;**162**:53-61

[77] Hoque AKMR, Aslam AFM, Ahmed M, Mamun MSA, Howlader AJ. Laboratory and field evaluation of an entomopathogenic fungus formulation-Bioterminator (Metarhizium anisopliae Metchnikoff) against termite infesting tea. Journal of Tea Science Research. 2016;**6**(9):1-6

[78] Fernandes PM, Alves SB. Control microbiano be Cornitermes cumulans with Beauveria bassiana and Metarhizium anisopliae under field conditions. Anais da Sociedade Entomológica do Brazil. 1991;**20**:45-50

[79] Razia M, Sivaramakrishnan S. Evaluation of entomopathogenic nematodes against termites. Journal of Entomology and Zoology Studies. 2016;**4**(4):324-327

[80] Shahina F, Tabassum KA. Virulence of Steinernema pakistanense against different insect species in laboratory condition. Pakistan Journal of Nematology. 2010;**28**:279-284

[81] Zaidawi JB, Karimi J, Moghadam EM. Entomopathogenic nematodes as potential biological control agents of subterranean termite, Microcerotermes diversus (Blattodea: Termitidae) in Iraq. Environmental Entomology. 2020;**49**(2):412-421

[82] Yu H, Gouge DH, Baker PB. Parasitism of subterranean termites (Isoptera: Rhinotermitidae) by entomopathogenic nematodes (Rhabditidae; Heterorhabditidae). Journal of Economic Entomology. 2006;**99**:1112-1119

[83] El-Bassiouny AR, El-Rahman RMA. Susceptibility of Egyptian subterranean termite to some entomopathogenic nematodes. Egyptian. Journal of Agricultural Research. 2011;**89**: 121-135

[84] Amarasinghe LD. Entomopathogenic nematodes from the costal belt of Sri Lanka and their efficacy in controlling termites. Pest Technology. 2008;**2**(2):125-129

[85] Singha D, Singha B, Dutta BK. In vitro pathogenicity of Bacillus thuringiensis against tea termites. Journal of Biological Control. 2010;**24**:279-281

[86] Devi KK, Kothamasi D. Pseudomonas fluorescens CHA0 can kill subterranean termite Odontotermes obesus by inhibiting cytochrome c oxidase of the termite respiratory chain. FEMS Microbiology Letters. 2009;**300**:195-200

[87] Osbrink WLA, Williams KS, Connick WJ, Wright MS, Lax AR. Virulence of bacteria associated with the Formosan subterranean termite (Isoptera: Rhinotermitidae) in New Orleans, LA. Environmental Entomology. 2001;**30**:443-448

[88] Myles TG. Isolation of Metarhizium anisopliae (Deuteromycotina: Hyphomycetes) from Reticulitermes flavipes (Isoptera: Rhinotermitidae) with convenient methods for its culture and collection of conidia. Sociobiology. 2002;**40**:257-262

[89] Federici BA, Bonning B, Leger RJ. Improvement of insect pathogens as insecticides through genetic engineering. In: Hill C, Sleator R, editors. Pathobiotechnology. Austin: Landes Bioscience; 2008. pp. 15-40

[90] Rath AC. The use of entomopathogenic fungi for control of termites. Biocontrol Science and Technology. 2000;**10**(5):563-581

[91] Milner RJ, Staples JA. Biological control of termites-results and experiences within a CSIRO project in Australia. Biocontrol Science and Technology. 1996;**6**:3-9

[92] Wright MS, Lax AR, Raina AK. Recent developments in the use of fungi for biological control of termites. In: Proceedings. XXII International Congress of Entomology, Brisbane. 2005

[93] Maniania NK, Ekesi S, Songa JM. Managing termites in maize with the entomopathogenic fungus Metarhizium anisopliae. International Journal of Tropical Insect Science. 2002;**22**:41-46

[94] Hussain A, Ahmed S, Shahid M. Laboratory and field evaluation of Metarhizium anisopliae var. anisopliae for controlling subterranean termites. Neotropical Entomology. 2011;**40**(2):244-250

[95] Intodia A, Prasad A, Veerwal B. Histopathology of Beauveria bassiana (Balsamo Vuillemin) an entomopathogenic fungus, infection in the midgut of termite, Odontotermes obesus (r.) worker. International Journal of Recent Scientific Research. 2019;**10**(8):34326-34330

[96] Singha D, Singha B, Dutta BK. Potential of Metarhizium anisopliae and Beauveria bassiana in the control of tea termite Microtermes obesi Holmgren in vitro and under field conditions. Journal of Pest Science. 2011;**84**:69-75

[97] Chouvenc T, Su NY, Grace JK. Fifty years of attempted biological control of termites – analysis of a failure. Biological Control. 2011;**59**:69-82

[98] Tamashiro M. Susceptibility of termites to microbes. A research proposal submitted to the Office of Naval Research. Department of Entomology, University of Hawaii, Honolulu; 1968. 7 p

[99] Wang C, Powell JE, Nguyen K. Laboratory evaluations of four entomopathogenic nematodes for control of subterranean termites (Isoptera: Rhinotermitidae). Environmental Entomology. 2002;**31**:381-387

[100] Qasim M, Lin Y, Fang D, Wang L. Termites and microbial biological control strategies. South Asia Journal of Multidisciplinary Studies. 2015;**1**(8):1-27

[101] Poinar GO. Nematodes for Biological Control of Insects. Boca Raton: CRC Press, Taylor and Francis; 1979. p. 249

[102] Georgis R, Poinar GO, Wilson AP. Susceptibility of dampwood termites and soil and wood-dwelling termites to the entomogenous nematode Neoplectana carpocapsae. International Research Communications Systems Medical Science. Microbiology Parasitology. Infectious Diseases. 1982;**10**:563

[103] Danthanarayana W, Vitarana SL. Control of the live wood tea termite. Glyptotermes dilatatus using

Heterorhabditis sp. (Nemat.). Agriculture Ecosystems and Environment. 1987;**19**:333-342

[104] Dolinski C, Lacey LA. Microbial control of arthropod pests of tropical tree fruits. Neotropical Entomology. 2007;**36**:161-179

[105] Smythe RV, Coppel HC. The susceptibility of Reticulitermes flavipes (Kollar) and other termite species to an experimental preparation of Bacillus thuringiensis Berliner. Journal of Invertebrate Pathology. 1965;**7**:423-426

[106] Omoya FO, Kelly BA. Variability of the potency of some selected entomopathogenic bacteria (Bacillus spp. and Serratia spp.) on termites, Macrotermes bellicose (Isoptera: Termitidae) after exposure to magnetic fields. International Journal of Tropical Insect. Science. 2014;**34**(2):98-105

[107] Peden D, Swaans K, Mpairwe D, Geleta N, Zziwa E, Mugerwa S, et al. Improving agricultural water productivity through integrated termite management. In: Wolde M, editor. Rainwater Management for Resilient Livelihoods in Ethiopia: Proceedings of the Nile Basin Development Challenge Science Meeting, Addis Ababa, Ethiopia, 9-10 July 2013. Nairobi, Kenya: ILRI; 2013

[108] Ramakrishnan R, Suiter DR, Nakatsu CH, Humber RA, Bennett GW. Imidacloprid-enchanced Reticulitermes flavipes (Isoptera: Rhinotermitidae) susceptibility to the entomopathogen Metarhizium anisopliae. Journal of Economic Entomology. 1999;**92**:1125-1132

[109] Yii JE, Bong CFJ, King JHP, Kadir J. Synergism of entomopathogenic fungus, Metarhizium anisopliae incorporated with fipronil against oil palm pest subterranean termite, Coptotermes curvignathus. Plant Protection Science. 2016;**52**:35-44

[110] Owusu EO, Akutse KS, Afreh-Nuamah K. Effect of some traditional plant components on the control of termites, Macrotermes spp (Isoptera: Termitidae). African Journal of Science and Technology. 2008;**9**:83-89

[111] Dokurugu M, Bernard BN, Irunuoh AO. Indigenous knowledge of termite control: A case study of five farming communities in Gushegu district of northern Ghan. Journal of Entomology and Nematology. 2012;**4**:58-64

[112] Nwilene FE, Agunbiade TA, Togola MA, Youm O, Ajayi O, Oikeh SO, et al. Efficacy of traditional practices and botanicals for the control of termites on rice at Ikenne, southwest Nigeria. International Journal of Tropical Insect Science. 2008b;**28**:37-44

Section 2

Hymenoptera

Chapter 4

The Paradox of the Old Queen Leaving

Jacques J.M. van Alphen

Abstract

Honeybees have a unique way of dispersal: the mother queen leaves with a swarm and settles elsewhere, while a daughter takes over the old nest. Studies of wild honeybees show that the probability of a swarm surviving its first winter is lower than that of an established nest. So, the departure of the mother queen seems paradoxical. I calculated the sum of the expected reproductive success of the mother and daughter for staying or leaving. The prediction is that the queen would have higher fitness returns from staying, due to the lower survival rate of swarms. I conclude that this cannot be the main reason why the old queen bee leaves. The departure of the old queen and the replacement by her daughter results in a drastically changed genetic composition of the colony. I suggest that the departure of the old queen is a way of giving the remaining colony a higher chance of surviving infectious diseases, and I provide the arguments for why this is the most likely explanation.

Keywords: honey bee, colony fissioning, swarming, reproductive success, inclusive fitness

1. Introduction

Most insects reproduce only once and then die. Eusocial insects are an exception to this rule. The queens of termites, ants, honeybees and stingless bees live for a number of years and reproduce over a number of seasons. Social insects have two fundamentally different modes of reproduction: independent colony foundation by single fertilised queens or colony foundation by fissioning of existing colonies into two or several new colonies (swarms). Reproduction by fissioning or swarm founding is found in ants, bees and wasps. It is found in most families of ants. In wasps, it is the characteristic reproductive mode of 24 polistine genera, as well as in some species of Ropalidia. In bees, it is found in the genus Apis and in stingless bees (Tribus: Meliponinae) [1].

As a rule, the queens of eusocial insects remain in the colony in which they started their reproductive life. In ants and termites, queens lose their wings and thus the ability to disperse. It is the fertile daughters of the queens that leave the nest and settle elsewhere to start a new colony. The exceptions to this rule are polygynous (with a number of queens) ant colonies and polygynous species of stingless bees [2]. Also, in many birds and mammals, the rule is that the mother stays at home, but the daughters leave for elsewhere [3], although there are some exceptions. In American red squirrels [4], Colombian ground squirrels [5] and kangaroo rats [6], some of the older

IntechOpen

Figure 1.
Swarm of A. mellifera *before moving to a new home (photo by the author).*

mothers disperse and leave the territory or the mound they lived in to their daughters, although the majority of mothers stay and the daughters leave. In social mammals whose daughters stay in the group, the mother also does not leave. The unique exception is the honeybees of the genus *Apis*. Honeybees reproduce *via* swarms that leave the old colony (**Figure 1**). The first and often the only swarm a colony produces leaves with the old queen, and a daughter takes over the old nest. Why does not the old queen just stay in the old nest, and why is not it a young queen that leaves with the swarm?

To answer that question, we need to investigate which is the best possible choice for the old queen. We must therefore calculate the expected reproductive success for that season for each of the two choices. The reproductive success of a queen is the sum of her current reproductive success through daughters and through sons, plus the queen's expected reproductive success in subsequent seasons (equal to the probability of her and her colony surviving the winter). This is known as inclusive fitness (IF) = direct fitness + indirect fitness [7, 8]. Inclusive fitness reasoning can be used to explain many characteristics of eusocial insects [9]. In the following, the discussion is limited to *Apis mellifera*, as it is the only species for which some data to calculate inclusive fitness are available.

2. Reproductive success through daughters

The reproductive success *via* daughters of a queen consists of the number of daughter queens that manage to become queen of a colony themselves, i.e., the number of swarms her colony produces. Usually, there is only one, and that is what we assume in the following. So, there is only one daughter who gets a colony. The other young queens die. The daughter has half the queen's genes; her relatedness is 0.5. Her contribution to the queen's fitness is therefore 0.5 * the probability that the young queen's colony survives into the next season.

3. Reproductive success through sons

The queen has produced several thousand drones with her colony. Only a few of these will succeed in fertilising a young queen, because there is huge competition between drones [10]. Even those drones that have succeeded in inseminating a queen are not sure that their genes will be used by the inseminated queen to fertilise the next generation of queens. This is because only a small number of young queens are produced while the queen has stored sperm from as many as 10–20 drones [11]. The probability of actually using a drone's sperm is thus less than 1. The contribution to reproductive success through sons thus comes a year later than that through daughters and therefore should be discounted for the delay. Since we do not know which drones succeed in mating and what exactly is the probability of their genes ending up in a young queen, we cannot properly calculate reproductive success *via* sons. We could get around this problem because we know that honeybees, on average, invest equally in the production of queens and drones [12]. This suggests that reproductive success *via* sons and daughters is equal. When the queen leaves with the swarm, she has already produced all the drones for that season. The same is true if she were to stay in the old nest. Since the reproductive success *via* sons is the same in both decisions, we can omit it from our calculations.

4. Future reproductive success

Because honeybee queens can reproduce for longer than one season, it is important for the queen's reproductive success that she and her colony survive the next winter. That survival rate probably depends on the decision to leave or stay in the old nest. The old nest has the advantage that there is no need to build combs, and food is already present. Leaving for a new home is a time- and energy-consuming process because the bees have to search for a new home and build new combs. This makes it more difficult for a swarm to gather enough food for winter stock. It is therefore expected that a swarm's chance of surviving until the next season is lower than for the old nest from which they left. It is not easy to find data to investigate this hypothesis. Because beekeepers do everything they can to help their bees survive the winter, we cannot use data on kept honeybees. Fortunately, two studies have measured these values for a population of wild honeybee colonies [13, 14]. In the forests around Ithaca, New York, the probability that an established colony was still there after winter was 0.80 in the 1970s and 0.84 in the 2010s. The probability of a swarm surviving its first winter was 0.29 in the 1970s and 0.23 in the 2010s.

5. The paradox

The queen must maximise her inclusive fitness and choose so that in the next season, there will be as many copies of her genes in the bee population as possible. Those copies are formed by the sum of her own genes plus half the genes of her daughter (the other half of the daughter's genes come from one of the drones that managed to mate with the old queen). Hence, her relatedness with her daughter is r = 0.5.

The sum to be maximised is then 1.0 times the probability that the old queen herself survives, plus 0.5 times the probability that the daughter survives. If we use the probabilities found by Seeley [10] in his study of wild colonies, we get for the 2017 data:

$$\text{When staying: IF} = (1 * 0.84) + (0.5 * 0.23) = 0.955 \tag{1}$$

$$\text{When leaving: IF} = (1 * 0.23) + (0.5 * 0.84) = 0.65 \tag{2}$$

So, the old queen would be better off staying at home and letting her daughter leave with the swarm; the same goes for the earlier 1970s experiment.

At this point, we conclude that a swarm's survival rate cannot be the main reason why the old queen bee leaves the nest with the first swarm. Only if the old queen and her colony had a lower survival rate if she stayed another year could the prediction be reversed. Most established colonies in Seeley's [3, 14] studies produced a swarm (95% in the 1970 experiment, 87% in the 2010 experiment). Those colonies thus had new queens. The question now is whether the survival of the established colonies would have been as high in case the old queen had stayed there.

The forests near Ithaca in New York may not be a very rich habitat for honeybees, and the winters there are cold and long. Would the outcome be different in a milder habitat? In a habitat where the probability of surviving the winter for a swarm is as high as the probability for the established colony, the values for leaving and staying are similar. Even then, leaving would not allow the old queen to increase her reproductive success any more than staying.

We now ask whether a daughter should stay or leave. The inclusive fitness for each of the decisions is:

$$\text{Staying: IF} = (1 * 0.84) + (0.25 * 0.23) = 0.8975 \tag{3}$$

$$\text{Leaving: IF} = (1 * 0.23) + (0.25 * 0.84) = 0.44 \tag{4}$$

Although the relatedness of a daughter queen to her sisters is 0.5 for sisters with the same father, most sisters have a different father, and therefore, the average relatedness is close to r = 0.25. So, a daughter queen would do better by staying. Given that the old queen leaves, we can conclude that no conflict exists between mother and daughter over the decision to stay or to leave. However, we have not answered the basic question of what the benefit is for the old queen to leave.

A different approach to this problem is asking if a new dominant mutant allele A coding for a queen to leave the colony after her first cycle of reproduction (mother queen leaves) would spread in the population. We first assume that mutant A occurs in an ancestral population with allele "a" fixed. The allele "a" codes for a queen to leave the colony before the first reproduction cycle (young queen leaves).

In a first scenario, the mutation occurs in one of the parents of the old queen and is passed on to 50% of her daughters. Some of these daughters will be raised as queens.

When the mother queen leaves with a swarm, she has a probability of surviving until the next season of only 0.23, but the original colony with copies of A benefits from her leaving with an increased survival probability (0.84). This allows the new allele A to spread in the population if the benefit to the original colony more than compensates for the loss of A copies as a result of the mother queen leaving. In the population that Seeley [13] studied, this is when the probability of survival is increased by >0.61. The behaviour of the mother queen can hence be considered as altruistic.

Now, consider a new dominant mutant allele A' coding for a queen to leave the native colony before her first cycle of reproduction (daughter queen leaves). A' occurs in a population with allele A fixed, with A coding for a queen to leave the colony after the first reproduction cycle (mother queen leaves).

A queen with the A' allele leaves the native colony with a swarm, thereby reducing her own chance of survival, with no beneficial effect on the colony from which she left, while there will be selection against the allele in the new colony.

We conclude that allele A can spread in a population and that a population in which A is fixed cannot be invaded by A'. This is only true when the departure of the old queen increases the survival probability of the old colony. The next question is then how the old colony may benefit from the old queen leaving.

6. What changes when the old queen's daughter takes over the nest?

The departure of the old queen and the takeover of the nest by a young queen have huge implications for the genetic composition of the colony. Although the young queen has half the genes of the old queen, she mates with 10–20 drones that are not related to the old queen, and this ensures that the workers in the original nest will come to have different fathers than if the old queen had stayed at home. If the old queen had stayed, the genetic composition of the colony's population would have remained the same. We could explain the departure of the old queen and the replacement by her daughter if the drastically changed genetic composition of the colony is important for the survival of the colony into the next season. This could be the case if, in a nest where the old queen remains, the likelihood of disease is higher than if the old queen leaves.

7. Why bees are vulnerable to pathogens and how they defend themselves

The high nest temperature, the many contacts between bees in a hive, and the frequent exchange of food mean that pathogens can spread rapidly in a colony. The first line of defence of honeybees against pathogens is that of antibiotic substances in plant resins collected by the bees to make propolis [15]. A natural bee's nest is surrounded by an envelope of propolis [16]. Since nectar and pollen also contain antibiotic substances, this food is at the same time a medicine chest [17]. Honeybees' main weapon and last line of defence consists of the immune system of individual bees.

8. The role of pathogens

Bacteria, fungi, microsporidia and viruses have much shorter generation times and populations many times larger than those of bees. As a result, they often produce new mutants, some of which could potentially break existing immune defences. The pathogens present in a colony can adapt to the colony in which they reside for an entire season. During the season, they had a large number of generations. In each of those generations, the pathogens that were best able to escape the bees' immune responses were at an advantage.

As long as the old queen is present, the genetic makeup of the bee colony does not change. But pathogens may continue to evolve and become more virulent during this period. Selection within a colony during such a relatively short period favours the more virulent pathogens because they produce more offspring than the less virulent variants. However, when the old queen leaves and is replaced by a young inseminated queen, new alleles enter the colony, potentially undoing the adaptation of pathogens to the colony and helping the bees to resist the pathogens present in the nest. The old queen migrates to a new environment, allowing her swarm to escape from the pathogens in the brood and stocks of the old nest. However, the old queen leaves with about 75% of the workers from the original colony [18, 19], and therefore in the new colony one would expect parasite load to be almost the same as and as detrimental as in the original colony, thus adding to its lower survival. Interestingly, small nests and frequent swarming lead to lower *Varroa* infestation and higher survival [20].

If the old queen were to stay, the pathogens could continue to adapt and become even more virulent for an entire season, threatening the survival of the colony. This hypothesis provides a possible explanation for why the old queen leaves. The prediction then is that if the old queen were to stay, a colony's chances of surviving the next winter would be lower than if a young queen took over. That prediction can be tested experimentally. There are currently no clear data known to me in the literature to support the hypothesis, although in a large experiment by Büchler et al. [21], 85% of the colonies died in the second season, and the survival curves dropped more quickly after the first year. In Seeley's [13, 14] studies, the odds of an established population producing a swarm were 0.95 in the 1970s and 0.87 in the 2010s. Seeley did not distinguish between established populations that produced a swarm and those that did not. But even if we knew the survival rates of colonies that did not swarm, we could not properly compare the two categories because of the low numbers of colonies that did not swarm and because colonies that do not produce a swarm differ from those that do. Colonies that do not produce a swarm are too small to do so or have insufficient stocks.

9. Discussion

Other species of the genus *Apis*, besides *A. mellifera*, live in tropical and subtropical regions. *A. mellifera* is the only species that lives in climatic zones with cold winters. The three main branches of the genus *Apis*—the giant honeybees, the dwarf honeybees, and the cavity-nesting honeybees—all produce swarms that leave with the old queen [18]. This suggests that this behaviour evolved before these species groups split and is a trait inherited from a common tropical ancestor. There are no published data on the survival rates of swarms of other *Apis* species. As swarms must invest time and resources in building a new nest and establishing food supplies, it is expected that their survival rates are lower than those of queens that take over the maternal nest. Therefore, the same paradox exists for the other Apis species.

I suggest that the hypothesis presented here—that the departure of the old queen is a way of giving the remaining population a higher chance of surviving infectious diseases—is the most likely explanation for this exceptional behaviour. With that, the old queen's departure would be one of a series of unique adaptations that honeybees have to protect themselves from pathogens. Those other unique adaptations are the mating behaviour that results in a panmictic population structure, the extreme polygyny of the queen [22], extremely strong sexual selection among drones, and the extremely high recombination frequency [23]. Young queens mate with as many as 10–20 different drones, and they use the sperm from all those drones to fertilise their eggs [11]. This ensures that the workers in a colony are the children of many different fathers and thus differ in heritable traits. This can limit the spread of pathogens in a population. The panmictic population structure and extremely high recombination frequency combine to form a double-edged sword in honeybees' fight against new bacterial or viral infections. Through mating behaviour, new rare alleles can be recruited from a large population, which can then be combined into new genotypes through recombination with useful alleles of other genes. In this way, honeybees can compensate for slower reproduction and lower population densities compared to those of bacteria and viruses.

That there is selection within a colony for increased virulence of pathogens does not contradict the theory that vertical transmission promotes the evolution of avirulence, while horizontal transmission (the spread of disease among unrelated individuals) promotes the evolution of virulence [24]. Selection for reduced virulence occurs over a longer time span through inter-colony selection: colonies with highly virulent pathogens will die, while those with less virulent pathogens can survive.

Preventing swarming, either by breaking the caps of young queens or by cutting off a wing of the queen, could therefore lead to a higher risk of infectious diseases for the colony.

Acknowledgements

JvA thanks Thom Seeley and Robin Owen for useful comments that improved an earlier version of the MS.

Author details

Jacques J.M. van Alphen
Naturalis Biodiversity Centre, Leiden, The Netherlands

*Address all correspondence to: jacques.vanalphen@naturalis.nl

References

[1] Hamilton WD. The genetical evolution of social behaviour I. Journal of Theoretical Biology. 1964;**7**:1-16

[2] Hamilton WD. The genetical evolution of social behaviour. II. Journal of Theoretical Biology. 1964;**7**:17-52

[3] Bourke AFG. Sex investment ratios in eusocial hymenoptera support inclusive fitness theory. Journal of Evolutionary Biology. 2015;**2**(8):2106-2111. DOI: 10.1111/jeb.12710

[4] Hovestadt T, Kohl P, Mitesser O. Optimal fissioning strategies of social insects with respect to colony dynamics and nest founding probability. Insectes Sociaux. 2024;**71**:171-183. DOI: 10.1007/s00040-024-00960-9

[5] Imperatriz-Fonseca VL, Zucchi R. Virgin queens in stingless bee (Apidae, Meliponinae) colonies: A review. Apidologie. 1995;**26**:231-244

[6] Clutton-brock TH, Lukas D. The evolution of social philopatry and dispersal in female mammals. Molecular Ecology. 2011;**21**:472-492. DOI: 10.1111/j.1365-294X.2011.05232.x

[7] Berteaux D, Boutin S. Breeding dispersal in female North-American red squirrels. Ecology. 2000;**81**:1311-1326. DOI: 10.1890/0012-9658(2000)081[1311,BDIFNA]2.0.CO;2

[8] Harris MA, Murie JO. Inheritance of nest sites in female Columbian ground squirrels. Behavioral Ecology and Sociobiology. 1984;**15**:97-102

[9] Waser PM, Jones WT. Survival and reproductive effort in banner-tailed kangaroo rats. Ecology. 1991;**72**:771-777

[10] Baudry E, Solignac M, Garnery L, Gries M, Cornuet JM, Koeniger N. Relatedness among honeybees (*Apis mellifera*) of a drone congregation. Proceedings: Biological Sciences. 1998;**265**(1409):2009-2014

[11] Page RE, Metcalf RA. Multiple mating, sperm utilization, and social evolution. The American Naturalist. 1982;**119**:263-281

[12] Page RE, Metcalf RA. A population investment sex ratio for the honey bee (*Apis mellifera* L.). The American Naturalist. 1984;**124**:680-702

[13] Seeley TD. Life history strategy of the honey bee, *Apis mellifera*. Oecologia. 1978;**32**:109-118

[14] Seeley TD. Life-history traits of wild honey bee colonies living in forests around Ithaca, NY, USA. Apidologie. 2017;**48**:743-754

[15] Simone-Finstrom M, Spivak M. Propolis and bee health: The natural history and significance of resin use by honey bees. Apidologie. 2010;**41**:295-311

[16] Seeley TD. The Lives of Bees: The Untold Story of the Honey Bee in the Wild. Princeton: Princeton University Press; 2019

[17] Erler S, Denner A, Bobis O, Forsgren E, Moritz RFA. Diversity of honey stores and their impact on pathogenic bacteria of the honeybee, *Apis mellifera*. Ecology and Evolution. 2014;**20**:3960-3967

[18] Seeley TD. Piping Hot Bees and Boisterous Buzz-Runners: 20 Mysteries of Honey Bee Behavior Solved. Princeton University Press: Princeton; 2025

[19] Rangel J, Seeley TD. Colony fissioning in honey bees: Size and significance of the swarm fraction. Insectes Sociaux. 2012;**59**:453-462. DOI: 10.1007/s00040-012-0239-5

[20] Loftus JC, Smith ML, Seeley TD. How honey bee colonies survive in the wild: Testing the importance of small nests and frequent swarming. PLoS One. 2016;**11**(3):e0150362. DOI: 10.1371/journal.pone.0150362

[21] Büchler R, Costa C, Hatjina F, Andonov S, Meixner MD, Le Conte Y, et al. The influence of genetic origin and its interaction with environmental effects on the survival of *Apis mellifera L.* colonies in Europe. Journal of Apicultural Research. 2014;**53**(2):205-214

[22] Tarpy DR, Seeley TD. Lower disease infection in honeybee (*Apis mellifera*) colonies headed by polyandrous vs monandrous queens. Naturwissenschaften. 2006;**93**:195-199. DOI: 10.1007/s00114-006-0091-4

[23] van Alphen JJM. Why honey bees are dying: A forgotten cause. Preprint. 2023. DOI: 10.20944/preprints202310.1544.v1

[24] Ray AM, Gordon AC, Seeley TD, Rasgon JL, Grozinger CM. Signatures of adaptive decreased virulence of deformed wing virus in an isolated population of wild honey bees (*Apis mellifera*). 2022. DOI: 10.1101/2022.12.09.519656

Chapter 5

Hymenoptera: Diseases, Parasites, and Pests

Akinwande Kayode Lawrence

Abstract

The decline in abundance and diversity of insects particularly social insects significantly impacts the ecosystem services most importantly pollination services.Other ecosystem functions impacted by this group of social insects that are crop pollinators, include nutrient cycle, food resources for animals and man and in bio-control of pests with some of the insects acting as parasitoids and parasites. The major sources of insect decline have been attributed to many biotic and abiotic stressors such as diseases, parasites and pests, habitat loss, alteration and fragmentation, pesticides, climate change, and invasive species. When social insects are simultaneously exposed to these multiple stressors, there are direct or indirect detrimental and negative interactive effects such as poor foraging and nutrition, suppression of immunity, and increase in disease susceptibility. In Social Hymenoptera, such as bees, ants, and wasps, the impact is more severe at both individual and colony levels because the insects are characterized by the presence of overlapping generations with cooperative social behavior. The litany of biotic stressors linked to the health of this insects are viruses, bacteria, fungi, protozoans, nematodes, mites, and pests.

Keywords: immunity, disease, decline, virus, colony

1. Introduction

1.1 Hymenoptera: do they get sick?

Hymenoptera is a unique order in the class insecta because of its services to the environment particularly in the terrestrial ecosystem where it provides important ecosystem services: pollination. The group also contributes to other functions provided by the general insect orders such as nutrient cycling, food (including for humans), control of pests (biocontrol) with some insects acting as parasitoids and parasites. Human perception about insects is that they are majorly vector of parasites and pathogens causing diseases of man, animals, and plants while some are destructive pests of our agricultural products. However, all insect orders are also victims of pests, parasites, and pathogens like other organisms. Even insects that are not incriminated to transmit pathogens and parasites and that are not pests are regarded as nuisance and aggressive because they sting, bite, and produce toxic and allergic reactions by injecting their venom into their victims. Some hymenopterous insects fall into this category; they sting or bite and produce venom consisting of biogenic amines, phospholipase, hyaluronidase, and other species-specific substances [1].

Diseases, parasites, and pests have been major factors in the decline of pollinating insects especially the hymenopterans. Although this single factor is not solely responsible, there are many other complex factors that interact together. These include habitat loss, alteration and fragmentation, pesticide use, climate change, invasive species, and others [2]. In the case of Honeybees, there is the global threat of Colony Collapse Disorder (CCD), an epidemic causing the rapid die-off of wild and managed bee colonies, the primary causes of which have been suspected to be pathogens and pesticides or the interactions between these biotic and abiotic stressors [3, 4]. On the other hand, the global spread of *Varroa mite*, *Varroa destructor*, and *Varroa jacobsoni* (Acari: Varroidea) is suspected to be directly connected to CCD. Varroa mite is an ectoparasite that feeds on the bee fat bodies and vector honeybee viruses to the colonies. Another global phenomenon is the growing spread of Small Hive Beetles. *Aethina tumida* (Coleoptera Nitidulidae), a kleptoparasite and pest of bee colonies, is native to sub-Saharan Africa [5, 6]. Both Small Hive Beetle and *Varroa mite* have been incriminated in the transmission of several honeybee viruses. Order Hymenoptera among the insect orders also encounters many challenges, the parasitoid fly, *Physocephala texana* (Diptera: Conopidae) and phorid flies, *Apocephalus borealis* (Diptera: Phoridae), a humpbacked fly attack bumblebees [7]. There is also a wide array of fungi; *Aspergillus, Paecilomyces, Metarhizium,* and *Beauveria* pathogens of Vespid Wasps; the bacteria *Serratia marcescens*, and *Bacillus thuringiensis*; and the nematodes *Heterorhabditis bacteriophora,* and *Steinernema* (= Neoaplectana) spp. also affect insects [8]. The emerging infectious diseases (EIDs) caused by viruses are killing insects, and recent findings show that honeybee viruses can be transmitted to other social insect species (bees, wasps, and ants) [9, 10].

The litany of pathogens, parasites, and pests associated with the order hymenoptera is responsible for the high mortality, and this definitely suggests the insects do get sick and the outcome of which is population decline.

2. Factors promoting the declining health of social insects

2.1 Eusociality

Eusociality in hymenoptera has predisposed the order of the insects to diseases, parasites, and pests. The high colony population, homeostatic regulation of colony nest temperature by most social insects increase the risk of disease outbreaks, growth and survival of pests and parasites [11, 12]. The cooperative feeding habit that involves processing and storing food in the nests by the social insects not only helps to maintain a continuous provision and abundance of food during bloom and dearth seasons for pests but also gives opportunities to many pests to come and inhabit permanently in their nests in search of food resources, despite the fact that most hymenopterans have developed a trophalaxy behavior to feed and share food among their caste members and exclude intruders [13]. However, many pests have developed a characteristic behavior that outsmarts this host habit [14].

The wax moths, a lepidopterous pest of honeybees, are attracted to the bee colonies by the food resources available in the nest comb. These wax moths include Lesser wax moth *Achroia grisella* and the Greater wax moth *Galleria mellonella* that feed on honey, honey combs, shed larvae skin, propolis, and pollen. The food resources enhance the moths' survival and reproduction, laying eggs inside the hidden corners of the beehives [14]. Similarly, the Small Hive Beetles (SHBs) (*Aethina tumida*), a small tiny beetle, native to sub-saharan Africa, are the beetles that primarily infest the nest of honeybees in search of food. In the nest, the beetle exhibits mimicry and trickery behavior for a stake of food from the nursing worker bees. The bees while taking food from the workers maintain contact with the mandibles of the workers presenting it. A similar habit is what the SHB mimicries; the pest allows its antennae and mouth parts or forelegs to touch the worker bees' mandibles. This trickery posture of this kleptoparasite SHB, which is a mimicry of honeybee trophallaxy, deceives the worker bees to offer it nourishing food, rich in carbohydrates and proteins; this food meets the needs of the beetles for survival and massive reproduction [13, 14]. Similarly, many parasitic flies that attack the honeybees, bumblebees, and wasps come to their nest in search of food.

Similarly, the *Apocephalus* Phorid flies, *Apocephalus borealis* that parasitizes bumblebees and paper wasps are also attracted to the insect nests by specific nesting materials [15] and the sweetness of their stored food [13, 14]. Furthermore, some biotic stressors, for example, fungi, bacteria, and viruses, take advantage of the bee social interactions and trophallaxy within the colony members of eusocial hymenoptera, the wasps, bees, and ants, to increase the risk of diseases by spreading them between infectious and susceptible individuals caste members, but this is unlikely in the solitary hymenopterous insects.

Many scientists believed that the vulnerability of hymenopterous social insects to infections is also connected to low genetic variability in their population, because all the colony members are from the same matrilineal, sharing the same genetic inheritance with little or no variation in their resistance or tolerance to diseases. However, it is still widely accepted that social insects should rather be more tolerant or resistant to diseases if eusociality is of any advantage because the insects' immune responses are a product of both individual organisms and the entire colony [16].

2.2 Foraging behavior

In hymenoptera, food foraging habit, plant that is foraged, and the quantity of food that is forage determine the insect survival, development, and reproduction. In an agro-environment, chemical applications for pest management and improving seed quality and quantity can adversely affect the forages and foraging insects [2, 17]. While the insects are foraging over large area, they encounter these abiotic stressors called toxic materials, which are of both natural and synthetic origin, and they bring these xenobiotics to the colony nests. In hymenoptera and other insect orders that commonly forage in chemically polluted agricultural and urban landscapes, there is continuous contact with the residues of these agrochemicals. The reaction of the insects on contact with these chemical residues and its severity is dependent on several factors: the type, dosage and duration of exposure with the chemical residues, and the synergism exhibited with other co-habiting factors such as nutritional and health status of the insects; all these are termed extrinsic factors [18].

The insect reaction to the xenobiotics could produce behavioral abnormalities, when it influences intrinsic and cognitive behavior. The impacts produced by xenobiotics, whether extrinsic and behavioral, influence the protective and defense mechanisms of the insects against infections and the infectious agents [19]. Findings have shown that synthetic insecticides used in agriculture have the tendency to increase the oxidative stress that could impair the production of some antimicrobial peptides in the insects; the chemicals could also damage the DNA and deform the cell cycle, thereby weakening the immune cells and rendering the organisms highly susceptible to infections and diseases that ordinarily would have been resisted or tolerated [16, 20]. The chemical can also cause cell death by inhibiting the proliferation of monocyte and lymphocyte cells leading to an inability to produce cytokinin and immunoglobulin. Therefore, scientists hold the opinion that social insects particularly the pollinators often exposed to xenobiotics are immunocompromised [19].

Malphighian tissues of social insects, on exposure to pesticides and phytochemicals in the environment and the flowers they visit, become activated for high expression of detoxification genes. CYP336 family of cytochrome enzymes within the P450 enzymes, a sister group to CYP3 clan (CYP9 and CYP6 families) commonly expressed in the Hymenoptera families Tenthredinidae, Vespidae, Apidae, and Formicidae to facilitate the detoxification of the phytochemicals and xenobiotics [20–23]. At the instance of this high expression of detoxification genes, there is a lowered expression of immune genes; thus social insects have fewer immune-related genes [20, 22]. Continuous encounter with toxic chemicals means continuous detoxification and relative weakening of immunity [23, 24]; therefore, a synergistic interaction exists between pesticides and pathogens. With exposure to xenobiotics, social insects become highly susceptible to pathogens and parasitic infections due to the weakening of their defense mechanism [20, 25].

Again, most hymenoptera exhibit some hygienic behavioral mechanisms such as grooming and uncapping and removal of dead and diseased larvae and pupae from sealed brood cells to prevent spread or reduce infections, parasites, and pest infestations [26]. There is loss of these intrinsic and cognitive memories of hygienic behavior to resist diseases and pests on exposure to pesticides poisoning. Thereby, the colonies remain susceptible and weak to pathogens and parasites with resulting manifestation of several symptomatic conditions.

Disappearance of plant resources due to habitat loss, deforestation, and agricultural practices has equally impacted the quantity and quality of forages. This problem has impaired the health of the insects. Where the insects have good nutrition, there is the tendency to mitigate the pesticide effects [2].

2.3 Climate change

Climate change is characterized by warming of the planet and changes in weather patterns. It is another significant factor contributing to the declining health and population of most insects. It also leads to decline in floral development, nectar and pollen production, alteration in the quality and quantity of the food nutrients, and asynchronous dynamics between flowering plants phenology and the foraging calendar. The aftermath is a reduction in foraging activities, poor development cycle of pollinator insects, and increased frequency of occurrence, spread, and diversity of diseases, pests, and parasites. This condition has manifested in the development of migration strategy for organisms to escape competition, predation, starvation, and invasive organisms and eventually created a great disturbance to the ecosystem.

Climate change negatively impacts the health of the ecosystems, with alteration in the distribution of plants, viruses, bacteria, animals, and even human settlements. Therefore, it brought about the risk of emerging diseases, pests, and parasites. With rising temperature, winter in the cold temperate regions begins to experience a raining phase; areas not known for hotness or dry summer and drought have begun to experience it [2]. Similarly, the tropics are becoming increasingly hot; globally, there is unusual flooding in areas where it is least expected. The surge in incidences of pests and parasites infestations originally restricted to the tropics, with no known symptomatic responses, has now become a global phenomenon with known symptomatic harms done to the hosts. Again, climate change has promoted invasive pests and parasites, with great virulence on their new hosts. Several organisms associated with the hymenoptera as pests and parasites belong to this category of these invasive animals. The SHBs *Aethina tumida*, parasitic mites *Tropilaelaps clarae*, Varroa mites *Varroa destructor* and *V. jacobsoni*, and protozoan parasites *Nosema ceranae* are included [5, 27, 28].

With the climate change, the increasing presence of carbon dioxide released through human activities is further promoting atmospheric warming. This situation tremendously impacts social insects as nest dwellers. When the nest ambient CO_2 level is high, the colonies tend to suffocate; in the leaf-cutter ants (*Atta cephalotes*), a fungus-farming ant, the ants exit the nest; otherwise inhabitants will be suffocated, their fungus garden will be affected, and the food supply may cease [29]. In honeybees and bumblebees, rising CO_2 levels in their nests lead to worker bees streaming outside the hive entrances for wing-fanning response to push fresh air through the hive until CO_2 levels return to normal, while other insects like ants that are wingless modify their nests by opening or closing entrances and exits to optimize ventilation [29, 30]. Although, no research work has detailed how the increased CO_2 level impacts the social insects' health, but the response of social insect to the change in CO_2 composition suggests it creates a physiological stress, and heat generated must be regulated; therefore, it decreases nest safety and probably increases the incidence of pests and diseases.

Three years of investigation (2016-2018) in Southwest Nigeria on climate change constraints/factors such as decrease precipitation which is characterized by low and shift in seasonal rainfall triggered late and heavy rainfall with heat waves, flooding etc. [2]. Several manifestations were recorded by Akinwande and Adeuya [2] in the area. The short time drought and the late appearance of heavy rainfalls witnessed heavy flooding that resulted in loss of arable land, wild forage plants, and bees forage crops. The flood also washed away some colony populations. During the period, the beekeepers recorded poor colony establishment, low honey and pollen yield, and weak colony health, survival, and colony density.

2.4 Environmental pollution

Environmental pollution is an abiotic stressor that has a great effect on insect health; it is the discharge of undesirable and unwanted organic and inorganic materials into the ecosystem due to anthropogenic activities and urbanization. The pollutants in the environment are pesticides, heavy metals, or airborne fine particulate matters, effluents, and discharges originating from agricultural or industrial sources. The release of these harmful substances leaves short- and long-term impacts on the environment, man, animals, and plants. Environmental pollution has contributed to decline in insect abundance and diversity globally, and the presence of the pollutants could also be linked to emerging deterioration in health of vertebrate and invertebrate animals; most important among the invertebrates are the insect pollinators. The consequence of pollutants on insects has far-reaching effects such as deterioration in pollinating activities and poor plant health, production, and yield.

Pollutants are known to elicit lethal or sublethal toxic effects. The social hymenopterans, such as bees, ants, and wasps, because of their activities to constantly exploit resources in the surroundings of their nest, agricultural, rural and urban landscapes, have become the victims of both lethal and sublethal health effects of these pollutants at the individual and colony levels. Many are dying due to the chronic effects of the pollutants, while the sublethal health effects of the pollutants cannot be overemphasized. The sublethal effects include increased susceptibility to disease, weak immunity, and decreased tolerance towards other environmental stressors.

The pollutants that the social insects encounter in their environment also interfere with the insect reproductive development, lifespan, immunity, neurophysiology, and fecundity. Therefore, the emerging decline in the health of social insects is a product of this environmental stressor called pollution.

3. Diversity of pests, pathogens, parasites, and parasitoids of social insects

Hymenoptera nesting habit facilitates the development of eusociality. Attracted to the nests are a myriad of organisms that are described as pests, pathogens, parasites, and parasitoids. The primary aims of their presence in the nests are to obtain food resources, for protective services provided the colonies, and to locate hymenoptera target hosts. For the pathogens, there are litany of organisms under this classification; these include viruses, bacteria, and fungi, while parasites include sporozoa, amoeba, ciliates, nematodes, trematodes, and cestodes; for pests, we have the coleopterans, dipterans, acarines, mollusks, diplopods and; parasitoids which are mainly flies and some wasps. All hymenopterans, both the wild and domesticated species, are threatened by the presence of these biotic stressors, with their virulence targeted at singular species-specific or a broad range of pollinator host or an entire colony. Pests and parasitoids of hymenoptera are not usually species specific [31] unlike pathogens such as viruses, bacteria, and parasites such as protozoans and helminths.

There are diverse species of pathogenic bacteria and viruses plaguing social and solitary hymenoptera. Bacteria are single-celled microscopic organisms, and viruses are basically genetic materials (RNA or DNA) encapsulated in a protective layer made of protein. Viruses can multiply only in living host cells [9, 10, 32]. Most viruses are super microscopic causing very different diseases, which appear similar in their symptoms. A parasite is an organism that lives on or in a host organism and gets its food from or at the expense of its host and can also change the behavior of the host.

Pest action is completely different from that of parasitoid. Pests are animals thatare noxious, destructive, nuisance, or troublesome to man or his interests, either animals or crops.

Pests of hymenoptera are found among insects, nematodes, mites, snails, slugs, and so forth and vertebrates like rats and birds. Some of the hymenoptera pests do not stay in their nests; some are conditional, and some cohabit with their hymenoptera host in the nests; however, those that cohabit with the host may have one or two stages in their life cycle in the nest and reproduce by interfacing with the outside environment, while some live their entire life in the host nests. Some of these pests exhibit some characteristic behavior to survive such as cleptobiosis (stealing the stored food for the colony), mimicry (posing a resemblance to the host), and hiding or having a protective coccon covering for their larvae. Some pests of these hymenoptera have thus been classified as kleptoparasitic and as potential vectors of pathogenic microorganisms.

Parasitoids are mostly insects; they develop at the expense of their hymenopterous host that they kill as a result of their interaction. Parasitoid reproductive success is dependent on their ability to successfully infest their host, and their survival and reproduction are dependent on the ability to overcome parasitism.

DOI: http://dx.doi.org/10.5772/intechopen.1005608

3.1 List of common pathogens, parasites, pests, and parasitoids of hymenoptera

3.1.1 Viruses of hymenoptera

Most viruses of hymenoptera are linked to Acarine mites particularly Varroa mites, and current research investigations have shown that pollinator hymenoptera, SHB *Aethina tumida*, *Vespula pensylvanica, Tropilaelaps mercedesae*, and *T. clareae* could also vector some of the viruses (**Tables 1** and **2**).

3.1.2 Pests and parasites of hymenoptera

3.1.2.1 Small hive beetle, a pest of global concern

The small hive beetle (SHB) belongs to holometabolous insect (Order: Coleoptera, Family: Nitidulidae). It is often referred to as a pest, scavenger, kleptoparasite, and

	Viruses	Classification	Hosts	Symptoms
1	*Solenopsis invicta* virus 1, 2, and 3 (SINV-1,-2,-3) [9, 10].	Dicistroviridae	Adult/worker fire ants	It alters the behavior, reduces foraging and sharing food, larva and queen become malnourished and die. Colony eventually collapses
2	Deformed wing virus (DWV) A, B, and C. [9, 10].	Iflaviridae	Honeybees, bumblebees, *Megachile* and *Colletes* bees Vespids, and other hymenoptera pollinators	It infects brood and adults in *Apis* resulting in poor wing development and early aging and death of adults
3	Israeli Acute bee Paralysis Virus (IAPV) [9, 10].	Dicistroviridae	Honeybees, *Megachile*, *Colletes* bees, and other hymenoptera pollinators	It slows down colony establishment and reduces offspring/brood production
4	Kashmir Bee paralysis Virus (KBV) [9, 10].	Dicistroviridae	Honeybees, *Megachile* *Colletes* bees, and other hymenoptera pollinators	Similar to IAPV
5	Acute Bee paralysis Virus (ABPV) [9, 10].	Dicistroviridae	Honeybees but rare in none *Apis*, but detected in Bombus spp.	It infects brood, workers, and adults in *honeybees* resulting in poor flight and low foraging, thoracic and abdominal hair loss, and low productivity and death
6	Black Queen Cell virus (BQCV) [9, 10].	Dicistroviridae	Honeybees	The virus is highly pathogenic causing queen mortality with the larvae or prepupal in the queen cell walls dark brown to black. Weak colony.
7	Chronic bee Paralysis Virus (CBPV) [9, 10].	Unclassified	Workers, drones, and queen honeybees and non *Apis* hymenoptera pollinators	It causes poor flight activity due to swollen/bloated abdomens, common in black hairless bees.
8	Lake Sinai Virus (LSV) [10, 33].	Sinhaliviridae	Honeybees	It causes colony mortality.

	Viruses	Classification	Hosts	Symptoms
9	Sacbrood Virus (SBV) [9, 10].	Iflaviridae	Brood and adults honeybee, *Megachile*, and *Colletes* bees.	Turn the insect cuticle leathery that it cannot molt and pupate. Larvae die and become dark, brittle, scaly, and easily removed from the cell with cell cap discolored, sunken, and perforated.
10	Varroa Destructor Virus (VDV-1 and VDV-2) [9, 10].	Iflaviridae	Honeybees, bumblebees, and other *Apis* pollinator	Young adult bees have deformed wings, high pupal mortality, reduced lifespan, overwinter mortality, and colony loss in most cases
11	Slow Bee Paralysis Virus (SBPV) [9, 10].	Iflaviridae	Honeybee, Bumblebees,	It causes anterior legs to paralyze within ten to twelve days of contact.
12	Bee Macula-like Virus (BeeMLV) [9, 10].	Tymoviridae	Honeybee	It resembles the pathology of VDV and other viruses transmitted by *Varroa*
13	Tobacco Ring Spot Virus (TRSV) [9, 10].	Tymoviridae	Honeybees	TRSV reduces bees' life; it affects guts, wings, antennae, blood, and all other body parts of bees.
14	Seco-like Virus [9, 10].	Secoviridae	Honeybees, bumblebee, and wild bees	Unknown

Table 1.
Some pathogenic and nonpathogenic viruses of hymenoptera.

	Causative organisms	Classification	Disease conditions	Hymenoptera
1	*Varroa mite* *Varroa detructor* *Varroa jacobsoni*	Acari: Varroidae	A condition known as varroasis (parasitic mite syndrome). It is a cosmopolitan, parasite of adults and larvae, feed on the fat body and haemolymph, cause anemia, colony collapse and transmitting viruses.	Honeybee, also found on flower-feeding bumblebees
2	*Acarapis woodi*	Acari: Tarsonemidae	Acaraposis. Low honey production, deformed wing and paralyzed bee, winter colony loss and slow build up in spring, death, dysentery.	Honeybee
3	*Locustacarus buchneri*	Acari: Podapolipidae	It infects the tracheal and respiratory tubes.	Bumblebees
4	*Acarapis woodii*	Acari: Tarsonemidae	It infects the tracheal and respiratory tubes, the mites create inability to fly, disjointed wings, and distended abdomens.	Honeybee
5	*Raymentia anomala* Womersley	Acari:Laelapidae	The mite is kleptoparasitic and its infestation could cause starvation in the bee colony.	Lasioglossum (Parasphecodes) spp. (Halictidae bees)

	Causative organisms	Classification	Disease conditions	Hymenoptera
6	*Tropilaelaps mercedesae* and *T. clareae*	Acari:Laelapidae	They weaken or kill brood or cause viral infections in surviving adults. There is loss protein in infested pupae and immunity decline.	Larva of *Apis dorsata* and *A laboriosa*, and rarely in *Apis mellifera*
7	*Galleria mellonella* (Greater wax moth) and *Achroia grisella* (Lesser wax moth)	Lepidoptera: Pyralidae	The moth eats the beeswax, larva of honeybees; pollen and feces released in the brood cells cause severe damage in the hive.	Honeybees
8	*Braula coeca* (Bee louse)	Diptera: Braulidae	The pests apparently do little harm.	Honeybee
9	*Nosema ceranae* and *Nosema apis*, now *Vairimorpha ceranae* and *V. apis,*	Microsporidian parasite Nosematidae: Dissociodihaplophasida	*N. apis* can aggravate dysentery, while *N. ceranae* does not have overt symptoms	Honeybees
10	*Xenos vesparum*	Strepsiptera: Xenidae	The reproductive health of parasitized male wasps is virtually unaffected.	Bees *and Paper wasps (Polistes wasp)*
11	*Steinernema hermaphroditum*	Nematoda: Steinernematidae	The mutually symbiotic pathogenic bacteria food source in the insect is released and it kills the insect.	Insects Ants
13	*Heterorhabditis elegans*	Nematoda: Heterorhabditidae	It kill the ants and is used as a biological control agent.	Insects: Ants
14	*Tetradonema solenopsis*	Nematoda: Tetradonematidae	It changes the body morphology	*Ants*
15	*Dicrocoelium dendriticum*	Platyhelminthes: Dicrocoeliidae	The worm enters the brain of the host and alter its behavior.	*Ants*
16	Mermithid nematode *Pheromermis* sp.	Nematoda: Mermithidae	It kills the host	Yellow jackets
17	Small hive beetles *Aethina tumida* [34, 35].	Coleoptera: Nitidulidae	It kills the brood, transmits viruses, and reduces honey productivity.	Honeybees, bumblebees, and other bee colonies

Table 2.
Common pests and parasites of hymenoptera.

sometimes as generalist. The adult SHB life cycle is spent in the nests of honeybees, bumblebees, and other nesting bees; there, it functions and exhibits life history strategies such as predation, host location, establishment of a habitat, trophallaxy, and reproduction [34].

The beetle reproduces by laying eggs in the comb; the eggs hatch into larvae that wander and migrate into the soil where the larvae pupate and emerge as adults that infest and re-infest the colonies. Adult small hive beetles (**Figure 1**) are light brown (w and w′), tan to reddish-brown (x and x′), dark- brown, and progressively darker (almost black) (y and y′) as their exoskeleton hardens, covered with fine hairs, and range in size from 5 to 7 mm (1/4″) in length, with females longer than males and both being oblong or oval in shape (**Tables 3** and **4**) [35].

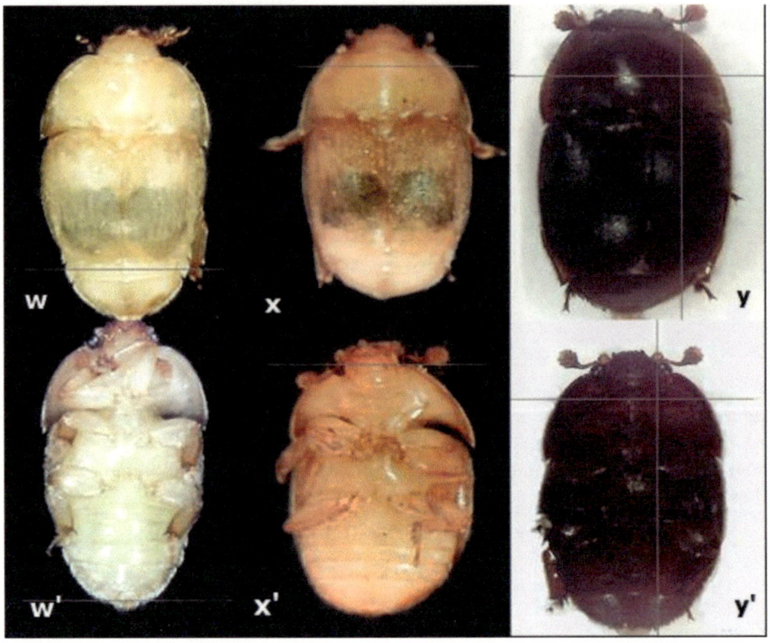

Figure 1.
Teneral adult (w) and adult (x and y) of small hive beetles. Teneral adult in preemergence stage from the soil showing w-dorsal and w′-ventral views; adult of small hive beetle in post emergence showing x-dorsal and x′-ventral views and y-dorsal and y′-ventral views (Akinwande et al. [34]).

3.1.3 Bacterial and fungal diseases of hymenoptera

	Bacteria and fungi	Diseases	Disease conditions	Hymenoptera
1	*Paenibacillus* bacteria spp	American foulbrood/ black brood/ropy brood disease	They are cosmopolitan, highly and rapidly infectious disease of prepupae	Honeybees
2	*Melissococcus pluton, Lactobacillus Eurydice, Paenibacillus alvei, Brevibacillus laterosporus, Enterococcus faecalis,* and *Paenibacillus apiarius*	European foulbrood	The queen produce scattered brood pattern showing the queen is sick, discolored, coiled, and rotting larvae with foul odor.	Honeybees
3	*Ascosphaera apis* (A fungus)	Chalkbrood or mummified disease	The queen produce scattered brood pattern showing the queen/colony is sick; the brood turn dry, black/ gray chalky mummies in the cells, hive entrance and bottom boards.	Honeybees
4	*Wolbachia* bacteria	Impaired reproduction	It affects reproduction in insects. This is referred to as reproductive parasitism	Ants and wasps
5	Ophiocordyceps unilateralis (Ascomycota: Hypocreales)	It causes death	The bacteria infect the ants, multiply in its tissues, and kill it	Ants

Table 3.
Bacterial and fungal pathogens of hymenoptera.

3.1.4 Parasitoids of hymenoptera

	Parasitoid	**Host**	**Impact**
1	*Vespa velutina* *Vespa mandarinia*	Honeybees	They invade the nest of the bees and feed on the bees.
2	Parasitoid wasp *Hybrizon buccatus* *Elasmosoma luxemburgense* *Neoneurus vesculus*	Ants	They lay eggs that develop to larvae in ants; the ants die, the larva use the ant to grow.
3	Conopid fly *Conops vesicularis* [7]	Bees and wasps	They lay eggs that develop to larvae that are endoparasites that kill the host
4	Phorid flies	Ant, Stingless and bumble bees	Phorid flies parasitoids on the host have larvae multiple fly larvae and end up killing the host from endoparasitism.

Table 4.
Common parasitoids of hymenoptera.

Author details

Akinwande Kayode Lawrence
Department of Biology, School of Life Sciences, Federal University of Technology, Akure, Nigeria

*Address all correspondence to: akinkay20032001@yahoo.com

References

[1] Everson GW. Hymenoptera. In: Wexler P, editor. Encyclopedia of Toxicology. 3rd ed. Vol. 2014. Amsterdam, London: Elsevier, Academic Press; 2014. pp. 984-985. DOI: 10.1016/B978-0-12-386454-3.00737-5. ISBN 9780123864550

[2] Lawrence AK, Adeuya EO. Nigeria experience of impacts of agricultural intensification and climate change on *Apis mellifera adansonii* colony establishment and health. Mellifera Journal. 2020;**20**(2):54-76. Available from: https://dergipark.org.tr/en/pub/mellifera/issue/27817/294461

[3] Boncristiani H, Underwood R, Schwarz R, Evans JD, Pettis J, van Engelsdorp D. Direct effect of acaricides on pathogen loads and gene expression levels in honey bees *Apis mellifera*. Journal of Insect Physiology. 2012;**58**:613-620

[4] Doublet V, Labarussias M, de Miranda JR, Moritz RF, Paxton RJ. Bees under stress: Sublethal doses of a neonicotinoid pesticide and pathogens interact to elevate honey bee mortality across the life cycle. Environmental Microbiology. 2015;**17**:969-983

[5] Lawrence AK, Neumann P. Small hive beetle infestation levels of honey bee colonies correlate with precipitation and forest cover. Apidologie. 2018;**49**:517-525. DOI: 10.1007/s13592-018-0579, Original article * INRA, DIB and Springer-Verlag France (2019) SAS, Part of Springer Nature

[6] Schäfer MO, Cardaio I, Cilia G, Cornelissen B, Crailsheim K, Formato G, et al. How to slow the global spread of small hive beetles, *Aethina tumida*. Small hive bettle contingency. Biological Invasions. 2019;**21**:1451-1459. DOI: 10.1007/s10530-019-01917-x

[7] Otterstatter MC, Troy LW, Robin EO. Contrasting frequencies of parasitism and host mortality among phorid and conopid parasitoids of bumble bees. Ecological Entomology. 2002;**27**:229-237

[8] Rose EAF, Harris RJ, Glare TR. Possible pathogens of social wasps (Hymenoptera: Vespidae) and their potential as biological control agents. New Zealand Journal of Zoology. 1999;**26**(3):179-190. DOI: 10.1080/03014223.1999.9518188

[9] Christian PD, Scotti PD. Advances in virus research. In: Encyclopedia of Virology. 3rd ed. Vol. 117. Amsterdam, London: Elsevier; 2023. pp. 2-136

[10] Anna G, Andrzej EM. Bee Keepers' Guide To Honey Bee Viruses. 2021st ed. USA: The Magazine of American Beekeeping; 2021

[11] Loope KJ, Wilson Rankin EE. Viral load, not food availability or temperature, predicts colony longevity in an invasive eusocial wasp with plastic life history. Scientific Reports. 2021;**11**:10087. DOI: 10.1038/s41598-021-89607-4

[12] Rowland BW, Rushton SP, Shirley MDF, et al. Identifying the climatic drivers of honey bee disease in England and Wales. Scientific Reports. 2021;**11**:21953. DOI: 10.1038/s41598-021-01495-w

[13] Neumann P, Naef J, Crailsheim K, Crewe RM, Pirk CWW. Hit-and-run trophallaxis of small hive beetles. Ecology and Evolution. 2015;**5**(23):5443-5743, Pages: i-iii

[14] Langlands Z, du Rand E, Crailsheim K, Yusuf A, Pirk Christian WW. Prisoners receive food

fit for a queen: Honeybees feed small hive beetles protein-rich glandular secretions through trophallaxis. The Journal of Experimental Biology. 2020;**224**:jeb.234807. DOI: 10.1242/jeb.234807

[15] Macivor JS, Salehi B. Bee species-specific nesting material attracts a generalist parasitoid: Implications for co-occurring bees in nest box enhancements. Environmental Entomology. 2014;**43**(4):1027-1033. DOI: 10.1603/EN13241. Epub 2014 Jun 23

[16] Evans JD. Immune pathways and defence mechanisms in honey bees *Apis mellifera*. Insect Molecular Biology. 2006;**15**:645-656. DOI: 10.1111/j.1365-2583.2006.00682.x

[17] Akinwande Kayode L, Anikwe JC, Adeonipekun PA, Okunrounmu T. Seasonal variation in forage plant, foraging time and duration for *Apis mellifera adansonii* in the forest and semi- savannah ecosystems of Nigeria. Mellifera Journal. 2016;**16**(2):1-12

[18] Yang E, Chuang Y, Chen Y, Chang L. Abnormal foraging induced by sublethal dosage of imidacloprid in the honey bee (Hymenoptera: Apidae). Journal of Economic Entomology. 2008;**101**:1743-1748

[19] Liao LH, Wu WY, Berenbaum MR. Behavioral responses of honey bees (*Apis mellifera*) to natural and synthetic xenobiotics in food. Scientific Reports. 2017;**7**(1):15924. DOI: 10.1038/s41598-017-15066-5

[20] Berenbaum M, Johnson RM. Xenobiotic detoxification pathways in honey bees. Current Opinion in Insect Science. 2015;**10**:51-58

[21] Adler LS. The ecological significance of toxic nectar. Oikos. 2000;**91**:409-420

[22] Hu YT, Wu TC, Yang EC, Wu PC, Lin PT, Wu YL. Regulation of genes related to immune signaling and detoxification in *Apis mellifera* by an inhibitor of histone deacetylation. Scientific Reports. 2017, 2017;**7**:41255. DOI: 10.1038/srep41255

[23] Stevenson PC, Nicolson SW, Wright GA. Plant secondary metabolites in nectar: Impacts on pollinators and ecological functions. Functional Ecology. 2017;**31**:65-75

[24] Sylvia CI, Christopher DP, Matthias AF. Social immunity: Emergence and evolution of colony-level disease protection. Annual Review of Entomology. 2018;**63**:105-123. DOI: 10.1146/annurev-ento-020117-043110

[25] Desneux N, Decourtye A, Delpuech JM. The sublethal effects of pesticides on beneficial arthropods. Annual Review of Entomology. 2007;**52**:81-106. DOI: 10.1146/annurev.ento.52.110405.091440

[26] Akinwande KL, Badejo MA, Ogbogu SS. Hygienic behavioral mechanism of resistance to diseases and parasitic infections in West African honey bee colonies *Apis mellifera* adansonii (Hymenoptera: Apidae) international. Journal of Entomological Research. 2014;**2**(2):73-79. Available from: https://esciencepress.net/journals/index.php/IJER/

[27] Akinwande KL, Badejo MA, Ogbogu SS. Challenges associated with the honey bee (*Apis mellifera* adansonii) colonies establishment in south western Nigeria. African Journal of Food Agriculture Nutrition and Development. 2013;**13**(2):1-18

[28] Akinwande Kayode L, Badejo MA, Ogbogu SS. Incidence of Korean

haplotype "K" of varroa mites infestation among bee colonies in Southwest Nigeria. Journal of Apicultural Research. 2012;**51**(4):369-370. DOI: 10.3896/IBRA.1.51.4.15

[29] Seeley TD. Atmospheric carbon dioxide regulation in honey-bee (*Apis mellifera*) colonies. Journal of Insect Physiology. 1974;**1974**(20):2301-2305

[30] Weidenmüller A, Kleineidam C, Tautz J. Collective control of nest climate parameters in bumblebee colonies. Animal Behaviour. 2002;**63**:1065-1071

[31] Heimpel GE, Abram PK, Brodeur J. A phylogenetic perspective on parasitoid host ranges with implications for biological control. Current Opinion in Insect Science. 2021;**44**:95-100. DOI: 10.1016/j.cois.2021.04.003

[32] Minhyeok K, Jung C, Eui-Joon K. Honey bee viruses: An ongoing history of discovery. Journal of Apiculture. 2023;**38**:121-137. DOI: 10.17519/apiculture. 2023.06.38.2.121

[33] Chen NC, Wang CH, Yoshimura M, et al. Structures of honeybee-infecting Lake Sinai virus reveal domain functions and capsid assembly with dynamic motions. Nature Communications. 2023;**14**:545. DOI: 10.1038/s41467-023-36235-3

[34] Lawrence AK, Mogaji OM, Alabi OA. Survival and development of the small hive beetle, *Aethina tumida* Murray (Coleoptera: Nitidulidae), in the soil. Bee World. 2020, 2020;**97**(3):90-95. DOI: 10.1080/0005772X.2020.1761207

[35] Ellis JD, Delaplane KS, Hood WM. Small hive beetle (*Aethina tumida* Murray) weight, gross biometry, and sex proportion at three locations in the southeastern United States. American Bee Journal. 2002;**142**(7):520-522

Chapter 6

Side Effects of Pesticides on *Trichogramma* Spp.

Sedighe Ashtari

Abstract

One of the most significant side effects of pesticides is their negative impact on natural enemies, which is evident in all agricultural ecosystems and is especially relevant in integrated pest control. While biocontrol alone is often insufficient to manage pests, the use of pesticides is unavoidable in many cases. Therefore, it is essential to be aware of the merits and demerits of pesticides and pay attention to their side effects on natural enemies. By implementing efficient management methods, the adverse effects of pesticides on natural enemies can be minimized. For this reason, it is necessary to study the effects of pesticides on natural enemies to enable their combined use as powerful pest management tools.

Keywords: *Trichogramma* spp., pesticides, side effects, IPM, IOBC

1. Introduction

Trichogramma spp. is a group of egg parasitoid wasps that serve as effective biological control agents for a wide range of agricultural and forest pests. There are at least 12 commercially used species of *Trichogramma* worldwide, making them a popular choice for biological control programs in various countries. These wasps have the advantage of being able to be reared on secondary hosts more quickly and easily compared to other biological control agents. Over the past few decades, different species of *Trichogramma* have been successfully employed in the biological control of pests such as the carob moth in pomegranate, rice stem borer, codling moth, European corn stem borer, and cotton bollworm.

Trichogramma wasps are classified under the genus *Trichogramma*, superfamily Chalcidoidea, and family Trichogrammatidae. They possess distinguishing characteristics that set them apart from other wasps in the superfamily. These include a smaller number of antenna segments (five to six), the presence of bristles on the wing edges, and a three-segmented tarsus. In contrast, other wasps typically have 10 to 12 antenna segments and a five-segmented tarsus. Within the Trichogrammatidae family, females are larger than males. Female *Trichogramma* have short, knee-shaped antennae with few and short hairs, a two-segmented funicle, and a clava without segments. In contrast, male *Trichogramma* have long antennae covered in numerous hairs, a funicle without segments, and a long clava without segments [1].

In the diagnosis of *Trichogramma* wasps, the morphological differences in the external male genital organs, as well as their internal and external sheaths, are utilized. These differences are used for species recognition and classification since there are clear variations among different species. Male and female *Trichogramma* wasps can be easily distinguished by their antennae. *Trichogramma* wasps exhibit an interesting reproductive behavior where they lay their oval and milky white eggs inside the eggs of other insects, particularly butterflies. These wasps have a life cycle that includes a free-living stage as complete insects (wasps), as well as internal parasitism in the eggs of other insects during the postembryonic period. They undergo stages of development, including larval, pre-pupal, and pupal stages. The larvae of *Trichogramma* wasps generally exhibit four distinct ages that can be differentiated to some extent based on their shape [2].

Trichogramma wasps undergo complete metamorphosis, exhibiting a holometabolous appearance. The larvae of these wasps have a broad and leaf-like shape. They feed on the contents of the host's egg, which is often found in the order of butterflies, particularly in families such as Pyralidae, Noctuidae, Pieridae, Tortricidae, and others. The larva then undergoes a transformation into a proto-pupa and eventually a pupa. Upon completing the pupation stage, the fully developed insect emerges from the host egg by creating a hole in the egg wall and immediately begins to fly. A notable characteristic during the development and parasitism of *Trichogramma* is the change in color of the host egg membrane, transitioning from its normal color to black. The number of eggs laid by *Trichogramma* in the eggs of pests varies depending on the size of the host egg. For instance, only one egg is laid on grain moth eggs, while two to four eggs are laid on *Heliothis* sp. (Lep: Noctuidae) eggs [1]. *Trichogramma* wasps are natural enemies of various agricultural, forest, and pasture pests and are found in most regions of the world. There are about 210 species of *Trichogramma* wasps, and at least 12 of these species are widely and commercially used in biological control programs [3]. Compared to other biological control agents, *Trichogramma* wasps can reproduce quickly, and their generation period is short. This makes them easier to breed on secondary hosts [3].

The effects of pesticides on natural enemies have been studied since the 1950s [4]. These studies have grown substantially since the mid-1970s, and many valuable materials have been published in this field over the past few decades [5]. Most conventional pesticides used for pest control are broad-spectrum organic compounds that can harm natural enemy populations in the same manner as pests, leading to a reduction in their numbers.

The reason for the lack of selectivity among conventional insecticides is that they target the nervous system, which is biochemically similar in both pests and natural enemies. This makes it challenging to find a truly selective combination of insecticides. Additionally, while the selected compounds may not have immediate effects on nontarget arthropods at recommended doses, prolonged exposure to small amounts of these chemicals can still have detrimental effects on their fitness [6].

According to researchers, pesticides have a side effect of reducing the population of natural enemies in agricultural ecosystems, particularly in areas with excessive pesticide use [7]. Studying the effects of pesticides in laboratory conditions helps identify pesticides that are compatible with biological control agents [8]. Insecticides disrupt the delicate balance between hosts and natural enemies. Multiple studies have demonstrated that insecticides used for pest control have various side effects on parasitoids, including alterations in growth and development, emergence percentage, and sex ratio [9].

DOI: http://dx.doi.org/10.5772/intechopen.1005089

The study of the side effects of pesticides on animal species in ecological systems is known as ecotoxicology. This field of science helps determine the compatibility of toxins with nature by examining their effects in these systems [10]. Selecting an insecticide that is relatively safe for natural and environmental enemies enables its use alongside biological control agents in integrated pest management programs. This approach ensures a more sustainable and environmentally friendly pest control strategy.

2. Side effects

2.1 Standard methods for evaluating side effects of pesticides on natural enemies

To assess the side effects of pesticides on natural enemies, the World Organization for Biological Control has developed a standard method. The results of this method, which is internationally accepted, provide valuable recommendations for integrated pest management programs [11]. The organization recognizes that a single type of test is insufficient to determine the impact of pesticides on beneficial organisms. Therefore, it has proposed three types of tests to comprehensively evaluate the side effects of pesticides on biological control agents. These tests aim to provide a more comprehensive understanding of the potential effects of pesticides on natural enemies.

2.1.1 The first test

The first test for evaluating the side effects of pesticides on natural enemies focuses on their critical life stages, such as the adult stage of *Trichogramma* wasps, which are most exposed to pesticides. The test involves calculating the LC50 amount, which is the amount of pesticide that causes 50% destruction of the tested arthropod population. Next, the risk ratio RQ (Ratio Quotients) is calculated both inside and outside the farm using the following equation: RQ = RD/LC50, where RQ is the risk ratio, RD is the recommended field dose (grams of effective substance per hectare), and LC50 is the lethal concentration of 50% based on milligrams of effective substance per liter.

2.1.2 The second test

The second test for evaluating the side effects of pesticides on natural enemies is performed on relatively resistant biological stages, such as those that pass inside the host's egg and are not exposed to pesticides much. In this test, researchers measure the mortality and side effects caused by pesticides on the reproduction of natural enemies using an index known as the Total Effect Index or E [12]. This method is a primary criterion for screening pesticides in the laboratory and is completed with field and semi-field tests. The World Organization for Biological Control, IOBC, has provided groupings based on the E index, which categorizes pesticides evaluated on natural enemies after testing in each of them (**Tables 1** and **2**) [10].

2.1.3 The third

The third test focuses on measuring the stability of pesticide compounds. Various factors such as sunlight, precipitation, soil reaction, humidity, soil organic matter, and

Mortality	Persistence	Category	Group
<30%	A, Short-lived	Harmless	I
30–79%	B, Slightly persistent	Slightly harmful	II
80–99%	C, Moderately persistent	Moderately harmful	III
99%	D, Persistent	Harmful	IV

Table 1.
Assessing of persistency and grouping of pesticides on natural enemies.

Insecticide	% Mortality				Classification
	3d	5d	16d	31d	
Spinosad	93.60 ± 0.73a	93.60 ± 0.73a	93.60 ± 0.73a	93.60 ± 0.73a	C, Moderately persistent
Indoxacarb	82.50 ± 0.72b	82.50 ± 0.72b	82.50 ± 0.72b	82.50 ± 0.72b	B, Slightly persistent
Imidaclopride	50.64 ± 0.74c	50.64 ± 0.74c	50.64 ± 0.74c	50.64 ± 0.74c	A, Short-lived

Table 2.
Mortality (%) of T. Brassicae *wasps after 24 hours exposure to residues of insecticides 3, 5, 16, and 31 days posttreatment (DPT) at laboratory conditions.*

heat can influence the stability and durability of pesticides, as well as their half-life in the environment. The distribution of pesticides in nature is primarily determined by their application method, formulation type, geographical factors, and weather conditions. The half-life of a pesticide in nature is categorized as short-lived if it is less than 30 days, moderate-lived if it falls between 30 and 100 days, and long-lived if it exceeds 100 days [13]. The duration of sunlight, type of pesticide, and radiation spectrum are crucial factors in the decomposition of pesticides on plant surfaces. This test provides insights into the persistence and potential environmental impact of pesticides.

In greenhouses with plastic cover, pesticides decompose faster than in those with glass cover due to the penetration of ultraviolet rays. The effect period of a pesticide can be determined by conducting stability tests. These tests are carried out in controlled environmental conditions by exposing beneficial insects to fresh and dried pesticide residues on plant organs like leaves. They aim to evaluate and determine the effect period of pesticides on nontarget arthropods. In field conditions, spraying is carried out and the treated product is protected from rain to prevent the pesticide residue from being washed away during the test period. The rain cover used in this test is permeable to ultraviolet rays. The final point of these studies is to determine the mortality rate after exposure to pesticide deposits for different periods up to 1 month after spraying [13, 14]. These studies provide valuable information on the persistence and potential impact of pesticides on nontarget organisms.

3. Effects of pesticides on natural enemies

The use of pesticides in agriculture is necessary to control pests and protect crops. However, the high consumption of pesticides has led to a decline in the population of natural enemies, resulting in the resurgence of target pests and the emergence of secondary pests in agricultural ecosystems worldwide. To address these issues, it is important to use pesticides that are selective for natural enemies, considering their ecological and physiological impact. Accurate knowledge of the effects of pesticides on pest populations and natural enemies is crucial for identifying suitable compounds. Evaluating the effects of insecticides on natural enemies should be comprehensive, taking into account both lethal and sublethal effects. This approach can help minimize the negative impact on natural enemies while effectively managing pest populations. Organic phosphorus compounds, carbamates, and pyrethroids are commonly used insecticides for pest control. However, these broad-spectrum insecticides can have adverse effects on natural enemies, particularly parasitoids. This is because a significant part of their life cycle involves searching for hosts, which is dependent on nerve transmission, the target site of nerve insecticides. As a result, they can cause higher mortality rates among natural enemies [15]. Pesticides have several side effects on pest control agents, including parasitoids. Exposure to pesticides can alter their lifespan (the time between exposure to the pesticide and the death of the agent), emergence rate, and parasitism rate. It is important to consider the impact of pesticides on natural enemies while managing pest populations to maintain ecological balance in agricultural ecosystems.

Pesticides can disrupt the precise coordination between an insect's nervous and hormonal systems, leading to behavioral and physiological disturbances related to egg-laying. The repellent properties of certain pesticides can indirectly disturb egg-laying behavior by reducing the chances of finding suitable hosts or suitable locations for egg-laying [16, 17]. Pesticides may also impact the sex ratio by affecting female gamete fertility, particularly in haploid species where female fertility can be regulated by the female insect. Additionally, the survival of natural enemies can be sex-dependent, especially when exposed to pesticides during their immature stage. These adverse effects can occur through direct contact with the pesticide or contamination of the host, leading to pest outbreaks due to poisoning and reduced effectiveness of control agents. To mitigate these effects, it is important to study the impact of different pesticide compounds on natural enemies and limit the use of compounds that have adverse side effects [18]. Understanding these effects can help develop more sustainable pest control strategies.

4. Methods of contamination of *Trichogramma* with pesticides

To investigate the side effects of pesticides on different life stages of parasitoid wasps, various methods can be used, including feeding whole insects with food impregnated with pesticides, direct contact with parasitized eggs of the host or whole insects with pesticides, or exposing whole insects to toxic residues. These methods have been used in different studies [19–21]. In field conditions, the eggs of parasitoid wasps are exposed to lower concentrations of insecticides, as they may be laid in shelters or under the leaves of plants, reducing their exposure to insecticides [22]. Understanding the effects of pesticides on different life stages of parasitoid wasps is crucial for developing effective pest management strategies that minimize the impact on natural enemies.

5. The effects of pesticides used on different species of *Trichogramma*

The use of insecticides can have detrimental effects on *Trichogramma* wasps, a parasitoid species. While the egg, larva, pre-pupa, and pupa stages of

Trichogramma are partially protected by the host's eggshell, insecticides can still disrupt both the mature and immature stages of the wasps. These disruptions include changes in the emergence rate of adults, parasitism rate, sex ratio, and lifespan of adult wasps across generations. Consequently, the performance of *Trichogramma* can be severely compromised [9, 23]. Most research on the effects of pesticides on pests and natural enemies has been conducted in laboratory conditions, primarily focusing on mortality as the evaluated parameter, without considering the conditions and ecological adaptation of surviving individuals. However, it is important to note that sublethal effects of pesticides can lead to physiological and behavioral changes in organisms. Insecticides can cause either lethal effects, resulting in organism death, or sublethal effects where other biological characteristics are altered without causing mortality. Understanding these effects is crucial for sustainable pest management strategies.

5.1 Effects of pesticides on *Trichogramma*

In their study, Souza et al. [24] highlighted the adverse impact of chemical pesticides on natural enemies in crops. These pesticides can cause lethal and sublethal effects, including death, disturbances in developmental rate, sex ratios, parasitism, and emergence rate [25]. However, research has shown that chlorantraniliprole has minimal harm on *Trichogramma japonicum* Ashmead. In fact, it was found to have the lowest mortality rate (21.25%) in adult *T. japonicum* [17]. Another study by Mahankuda et al. [26] indicated that chlorantraniliprole was slightly harmful to adults of *T. chilonis*. Field tests conducted on *T. chilonis* adults under semi-field conditions revealed that chlorantraniliprole had no harmful effects on this species and was classified as a short-lived persistence pesticide [27]. However, it is important to note that the recommended field concentration of chlorantraniliprole was found to be slightly harmful to adults of *T. chilonis*, *T. pretiosum*, and *T. japonicum* [28].

In a laboratory experiment conducted by Mohapatra and Shinde [29], chlorantraniliprole was found to be effective for female adults of *Trichogramma japonicum*. Another compound called lufenuron, which inhibits chitin production, was studied by Carmo et al. [30]. When lufenuron was applied to *T. pretiosum* Riley during the pupal stage, it resulted in a reduction in the emergence of female adults in the F1 generation [31]. A similar study conducted by Fontes et al. [32] also explored this topic. A study conducted on immature stages of *T. achaeae* investigated the effects of several insecticides, including lufenuron, under laboratory conditions. The results indicated that lufenuron had low toxicity in all developmental stages [33]. However, another study found that lufenuron was classified as a dangerous insecticide [33]. Afshari et al. [34] reported few side effects of lufenuron on adult and prepupae stages of *T. brassicae*. In contrast, the effect of abamectin was evaluated by Hussain et al. [35] on *T. chilonis* under laboratory conditions and was found to have adverse effects on the emergence rate of adults. In a study conducted by Khan and Ruberson [36], the effect of the field recommended concentration of abamectin on larval and pupal stages of *T. pretiosum* was examined. The results showed that abamectin significantly decreased the emergence percentage of adult parasitoids. Carvalho et al. [31] reported that acetamiprid had slight harmful effects on the developmental stages *of T. pretiosum*. Hewa-Kapuge et al. [37] found that emamectin benzoate was moderately persistent for *T. chilonis*. Another study by Mollanorouzi et al. [38] investigated the effect of two insecticides, indoxacarb and thiodicarb, on the mortality rate of *T. brassicae* adults. The results showed that indoxacarb caused less mortality compared to thiodicarb. In a study by Vaez et al. [39], it was found that the biological parameters of *T. brassicae* treated with a 20% concentration of lethal *Bacillus thuringiensis* were not affected.

In a study by Tabebordbar et al. [40], the effects of deltamethrin, spirotetramat, and flupyradifurone on different immature stages of *T. evanescens* were investigated. The results showed that deltamethrin had a negative impact on the biological parameters of larva, pre-pupa, and pupa stages. On the other hand, spirotetramat and flupyradifurone exhibited relative compatibility with the wasp, with the pupa stage being the most sensitive to these compounds.

In another study by Shirazi et al. [41], the recommended concentrations of diazinon and neem were evaluated in laboratory conditions on *T. brassicae* adults and parasitized eggs of *Sitotroga cerealella*. It was found that diazinon was effective for the emergence of immature stages but harmful for the survival of adults, classified as toxicity group 4. Neem, on the other hand, was found to be harmless for all stages and placed in toxicity group 1.

In a study conducted by Ashtari et al. [42], the effect of 25% lethal concentration of abamectin, emamectin benzoate, acetamiprid, and flubendiamide pesticides on *T. brassicae* and *T. evanescens* was investigated. The highest and lowest mortality percentages after 24 hours were observed in the abamectin and flubendiamide treatments. Another study by Ashtari [43] examined the effect of LC25 of spinosad, imidacloprid, indoxacarb, and chlorantraniliprole on the adults of *T. brassicae* and *T. evanescens*. The mortality percentages for *T. brassicae* were 31.8, 26.6, 25.6, and 23% respectively, while for *T. evanescens,* the percentages were 33, 28, 25, and 24% respectively. The results indicated that chlorantraniliprole had lower toxicity for parasitoid wasps compared to the other insecticides investigated. According to Parsaeyan et al. [44], treating the immature stages of *T. chilonis* Ishii and *T. brassicae* with the recommended field dose of emamectin benzoate negatively affects the lifespan of adults in the next generation. In a study by Ashtari et al. [45] on *T. evanescens*, the effects of field recommended concentrations and reduced doses of acetamiprid, abamectin, emamectin benzoate, and flubendiamide on the immature larval, pre-pupa, and pupa stages were investigated. Abamectin, in both concentrations, was found to be harmful in the lifespan test. Flubendiamide had a lesser impact on the lifespan of female adults compared to other insecticides. Shafaghi et al. [46] examined the effect of chlorantraniliprole, indoxacarb, and flubendiamide on *T. brassicae*. The results indicated that chlorantraniliprole had the least effect on the emergence of *Trichogramma* adults from treated eggs.

According to Parsaeyan et al. [47], treating the immature stages of different species of *T. chilonis* and *T. brassicae* with the recommended field dose of emamectin benzoate has a harmful effect on the reproductive rate of adults. In a study by Saber et al. [48], the effect of fipronil and diazinon on the functional reaction of *T. brassicae* was investigated. The researchers concluded that the functional reaction of bees in the control and fipronil treatment was of the second type, while in the diazinon treatment, it was of the third type. The lowest time of achievement was observed in the diazinon treatment. According to Bhargavi [49], studies conducted on the effect of emamectin benzoate on *T. chilonis* in laboratory conditions found that it is possible to release this wasp by observing a time interval of 4–5 days after the application of the pesticide. In another study, researchers determined the stability of recommended field concentrations of three insecticides, spinosad, imidacloprid, and indoxacarb, in semi-field conditions using the foliar spraying method [50].

According to Ashtari et al. [51], leaves of treated tomato plants were sampled and analyzed at different time intervals after application of insecticides. Acetamiprid and flubendiamide were classified as short-lived insecticides, causing mortality of 30.5

and 31.6% for *T. brassicae* and *T. evanescens*, respectively. Similarly, abamectin showed slight persistence with mortality rates of 16.1 and 13.8% within 16 days. Emamectin benzoate was moderately persistent, causing mortality rates of 13.3 and 15.5% within 30 days. Therefore, flubendiamide and acetamiprid were considered non-harmful to both *T. brassicae* and *T. evanescens*, making them suitable candidates for integration into IPM programs along with biological agents. According to Nekunam et al. [33], a study investigated the persistence of four insecticides (thiodicarb, indoxacarb, lufenuron, and proclim fit) on cotton plants and their toxicity to the parasitoid *T. brassicae*. Thiodicarb was classified as moderately persistent, while indoxacarb and proclim fit were classified as short-lived insecticides. Lufenuron was considered nonpersistent. In another study by Ashtari [52], the persistence of four insecticides (tetraniliprole, chlorantraniliprole, lufenuron, and thiocyclam) on tomato plants and their toxicity to two species of parasitoid wasps (*T. brassicae* and *T. evanescens*) were investigated. Thiocyclam exhibited longer stability and was classified as a moderately persistent insecticide. Tetraniliprole, chlorantraniliprole, and lufenuron were classified as short-lived insecticides. According to Alsaedi et al. [53], an investigation was conducted to assess the efficiency of three *Trichogramma* species (*T. brassicae*, *T. embryophagum,* and *T. evanescens*) on *Tuta absoluta* eggs treated with *B. thuringiensis* Kurstaki. The results showed that *T. embryophagum* had the highest parasitism rate at 54%, while *T. evanescens* had the lowest rate at 25%. *T. brassicae* exhibited a parasitism rate of 40%. The highest mortality percentage was observed in *T. evanescens* at 85.85%, while the lowest was in *T. embryophagum* at 37%. The mortality rate for *T. brassicae* was estimated to be 39%.

According to Vaez et al. [39], a study investigated the impact of *B. thuringiensis* on the biological parameters of *T. brassicae*. The results indicated that this treatment did not affect the reproductive capacity, emergence, and survival of the parasitoid. In another investigation by Nozad-Bonab et al. [54], the side effects of the LC50 concentration of *Metarhizium anisopliae* on the immature stages of *T. brassicae* inside *Tuta absoluta* eggs were examined. The findings revealed that this fungus did not have a significant impact on the biological parameters of *Trichogramma.* Additionally, Jafari et al. [55] studied the effect of the LC30 concentration of three essential oils (marjoram, basil, and mint) on the percentage of parasitism of *T. brassicae*. The highest rate of parasitism on *Phthorimaea operculella*, after the control treatment, was recorded as 64.44% for basil essential oil.

According to Rafiqi et al. [56], the effect of two insecticides, imidacloprid and hexaflumuron, on the hatching percentage of *T. brassicae* eggs, was investigated by immersing parasitized *Sitotroga cerealella* eggs in various concentrations. The findings revealed that hexaflumuron was less toxic than imidacloprid. When using *Trichogramma*, a minimum time interval of 72 to 96 hours after release should be considered. In another study by Shirazi et al. [41], field recommended concentrations of neem were examined in laboratory conditions on *T. brassicae* adults and parasitized eggs of parasitized *Sitotroga cerealella* on days 1, 3, and 7 after parasitism. The results showed that neem was harmless for all stages and placed in group 1. Moreover, Jafarinodooshan et al. [57] investigated the effect of Anghuzeh extract on the rate of parasitism and mortality of *T. brassicae* and the percentage of parasitism of eggs *Ectomyelois ceratoniae* Zeller. The findings indicated that Anghuzeh extract increased parasitism.

According to Khorrami et al. [58], a study was conducted to investigate the effect of catnip extract on the rate of parasitism of *T. brassicae*. The results showed that catnip extract did not have any negative impact on the rate of parasitism, which was estimated at 66.66%. In another investigation by Alimirzanejad et al. [59], the effect of four fungicides, carpropamide, tricyclazole, propiconazole, and mancozeb, was studied on eggs parasitized by *T. brassicae* in laboratory conditions. The findings revealed that these fungicides had harmful effects on *Trichogramma* wasps, contrary to the initial assumption. None of the fungicides were harmless to *Trichogramma,* and propiconazole had the most negative impact on the rate of emergence of treated eggs.

In a recent study conducted by Shirazi et al. [41], the effects of edifenphos fungicide on adults and parasitized eggs of *Sitotroga cerealella* by T. *brassicae* have been investigated under laboratory conditions. The recommended concentrations of the fungicide were tested on days 1, 3, and 7 after parasitism. The results revealed that the fungicide was found to be harmless for the immature stages of the insects. Furthermore, the study also examined the impact of butachlor and oxadiargil herbicides on adults of *T. brassicae* and parasitized eggs of *Sitotroga cerealella* on the 1st, 3rd, and 7th days after parasitism. It was observed that both herbicides exhibited toxicity toward the adult *T. brassicae*, placing them in group 4 according to **Table 1**. However, these compounds were found to be harmless to the immature stages, categorizing them as group 1 in **Table 1**.

In a recent study, the effects of several compounds on *T. pretiosum* have been examined. The results indicated that teflubenzuron had no significant impact on female survival and showed minimal effects on *T. pretiosum* in both direct-spraying and ingestion bioassays. In the residue contact bioassay, teflubenzuron was considered harmless, while flupyradifurone was rated slightly harmful. On the other hand, chlorfenapyr, thiodicarb, and methomyl were found to be harmful to the parasitoid through all exposure routes and exhibited persistence [60].

In a separate study by Parreira et al. [61], it was suggested that essential oils of *Mentha piperita*, *Piper nigrum*, and *Thymus vulgaris* could be incorporated into integrated pest management programs for *Trichogramma galloi*.

The lower side effects of these oils on the F2 generation of *Trichogramma galloi* indicate that gradually, the tolerance of wasps to these substances increases, which can pave the way for controlling some pests by simultaneously using *Trichogramma* wasps and these oils. Even by conducting more research projects on this species and other species of *Trichogramma* wasps, the reasons for the development of these mechanisms can be investigated and strengthened.

According to a recent project, the effects of various compounds on *T. pretiosum* have been evaluated. Thiamethoxam was found to be harmless, while abamectin, chlorfenapyr, and spinetoram were slightly toxic. Azadirachtin and lambda-cyhalothrin were moderately toxic, and chlorpyriphos was classified as toxic. Moreover, the emergence rate of the second generation was not significantly affected by thiamethoxam, abamectin, azadirachtin, and chlorfenapyr. However, sublethal effects were observed in the second generation due to azadirachtin, abamectin, and chlorfenapyr [62].

6. Reducing the side effects of pesticides

Chemical control is widely used in agricultural pest management, but its improper use can lead to various negative consequences such as environmental pollution, secondary pest outbreaks, resistance development, risks to human health, and harm to nontarget organisms. To mitigate these side effects, several measures can be implemented:

1. Integrated Pest Management (IPM): Adopting an integrated approach that combines various pest control methods, including biological control, cultural practices, and the targeted use of pesticides.

2. Proper Application Techniques: Following recommended guidelines for pesticide application, such as using the correct dosage, timing, and application methods to minimize off-target effects.

3. Selective Pesticides: Choosing pesticides that have minimal impact on nontarget organisms while effectively controlling the target pests.

4. Monitoring and Thresholds: Regularly monitoring pest populations and establishing action thresholds to determine when pesticide intervention is necessary, avoiding unnecessary applications.

5. Education and Training: Providing farmers and applicators with proper training and knowledge on pesticide handling, safety precautions, and integrated pest management practices.

By implementing these measures, the side effects of pesticides can be reduced, promoting more sustainable and environmentally friendly pest management practices in agriculture. The presence of weeds and trees in and around fields can provide shelter for natural pest control agents. It is beneficial to avoid removing all weeds, particularly at the field edges, as they can serve as a habitat for these beneficial organisms. In some cases, pests themselves, like aphids that cause leaf curling, can create shelters for their natural enemies. By preserving these habitats, the natural enemies can thrive and remain unaffected by pesticide exposure. This approach supports the conservation of beneficial insects and promotes natural pest control in agricultural ecosystems.

6.1 Wind breaker

Planting rows of trees or shrubs on the sides of farms can serve as windbreaks to prevent pesticides from drifting beyond the targeted field. The use of windbreaks is an effective method to minimize the off-target effects of pesticides, particularly in areas with high wind intensity. The trees or shrubs act as a physical barrier, reducing the wind speed and creating a sheltered area for the pesticide application. This approach can help to protect neighboring crops, natural habitats, and water sources from pesticide contamination, promoting more sustainable and environmentally friendly pest management practices in agriculture.

6.2 Attention to the level of economic loss

Continuous pest monitoring and accurate identification of insects through sampling and counting can help determine the economic damage threshold. By initiating control measures at the appropriate time, the cost of spraying can be reduced, and further damage can be prevented.

In addition, employing nonchemical methods such as proper plowing can disrupt pest life stages while promoting water and soil conservation. Weeding also contributes to the reduction of pest populations. It is important to exercise caution when using nitrogenous fertilizers, as excessive use can lead to outbreaks of sucking pests. Furthermore, utilizing resistant or tolerant cultivars can mitigate the damage caused by pests, while winter water ice can eliminate certain stages of their life cycle. These approaches offer solutions to reduce pesticide usage and promote sustainable pest management practices.

6.3 Spot spraying

Continuous monitoring of pest infestation and identifying infestation centers allows for targeted spot spraying or spraying until the pests are contained within the field. This approach helps to reduce pesticide usage and minimize associated risks. Selecting the appropriate combination and dose of pesticides is crucial in minimizing the impact on natural enemies. Factors such as the type of pesticide, formulation, dosage, and application method can significantly affect the sensitivity of beneficial organisms. For instance, systemic pesticides have a lesser impact on natural enemies compared to sucking pests. Additionally, it has been observed that digestive pesticides have a lower effect on beneficial agents when compared to contact compounds. By considering these factors and making informed choices, it is possible to optimize pest control while minimizing harm to beneficial organisms, promoting a more balanced and sustainable approach to pest management.

6.4 Using the right formulation

Microcapsule, granule, and pellet formulations are considered less harmful to beneficial organisms that are highly mobile. These formulations exhibit reduced loss compared to round formulations, which are more prone to be searched by insectivores. However, poisoned baits are considered the safest type of formulation for pest control while preserving natural enemies. By using poisoned baits, pests can be effectively targeted while minimizing the direct exposure of beneficial organisms to harmful pesticides. Considering the formulation type is important in order to strike a balance between effective pest control and the preservation of beneficial organisms. Choosing formulations that are less harmful to mobile beneficial organisms and utilizing poisoned baits can contribute to a more sustainable approach to pest management.

6.5 How to use pesticides?

The method of pesticide application plays a crucial role in minimizing the impact on nontarget organisms. Aerial application of pesticides, in particular, increases the risk due to high drift rates. It is important to carefully adjust the spraying time to minimize the exposure of nontarget organisms. Ideally, pesticide compounds should be used during the most sensitive biological stage of the pest and the most resistant stage of the natural enemy, if possible. This targeted approach can help protect natural enemies while effectively controlling pests. Furthermore, using reduced doses of pesticides can have a significant positive effect on the preservation of natural enemies. By minimizing pesticide exposure, the beneficial organisms can continue their role in pest control.

By considering these factors and implementing appropriate application methods, it is possible to mitigate the impact on nontarget organisms and promote a more sustainable approach to pest management.

By using pesticides with lower shelf life and environmental stability, adjusting spraying techniques, and incorporating electrostatic devices, bubble nozzles, and hallway sprayers, we can effectively reduce the impact of pesticides on natural enemies. These strategies not only preserve consumer health but also provide better protection and support for beneficial organisms [63].

7. Conclusion

When evaluating the effects of pesticides on natural enemies, it is crucial to consider both their lethal and sublethal effects. Sublethal concentrations of pesticides can have varying impacts on biological parameters, which can sometimes be more significant than their lethal effects. However, it is worth noting that most studies on the side effects of pesticides on *Trichogramma* species have been conducted in laboratory conditions, which may not accurately reflect real-world scenarios. To obtain more practical results, it is advisable to investigate the toxicity of new pesticides in greenhouse or semi-field conditions. Additionally, since pests and natural enemies coexist in natural environments, it is important to assess the combined effect of pesticides on both the pest and the natural enemy, allowing for necessary comparisons.

However, it is worth noting that most of the studies have primarily focused on insecticides, with less attention given to herbicides, fungicides, acaricides, biological, and herbal products. Additionally, the simultaneous effects of multiple insecticides or the combination of an insecticide with a fungicide have not been thoroughly investigated. Considering these aspects would lead to more realistic and comprehensive results. Moreover, there is a lack of experiments related to pesticide stability within the country, except for a few isolated cases. It appears that environmentally compatible pesticides like *Bt*, neem, marjoram, basil, and so on hold great potential for integrated pest management and effective control. These pesticides have shown minimal harmful effects on both the mature and immature stages of natural enemies such as mint, cat grass, and anguze. Based on studies examining the impact of plant and microbial toxins on *Trichogramma*, these compounds are considered a viable alternative to chemical toxins. The use of pesticides with lower adverse effects not only mitigates environmental consequences but also offers enhanced protection for natural enemies. Additionally, it safeguards consumers of agricultural products from any undesirable side effects associated with these compounds.

Conflict of interest

The authors declare no conflict of interest.

DOI: http://dx.doi.org/10.5772/intechopen.1005089

Author details

Sedighe Ashtari
Plant Protection Research Department, Markazi Agricultural and Natural Resources Research and Education Center, Agricultural Research, Education and Extension Organization, AREEO, Arak, Iran

*Address all correspondence to: aroya95@gmail.com

References

[1] Knutson A. The Trichogramma Manual. USA: The Texas A & M university system; 2005. Available from: http://insects.tamu.edu/extension/bulletins/b-6071.html

[2] Pinto JD. A review of the new world genera of Trichogrammatidae (hymenoptera). Journal Hymenoptera Research. 2006;**15**:38-163

[3] Consoli FL, Parra JRP, Zucchi RA. Egg parasitoids in agroecosystems with emphasis on *Trichogramma*, progress in Biological Control. 1st ed. Vol. 9. New York, USA: Springer; 2010. 482 pp

[4] Theiling KM, Croft BA. Pesticide side-effect on arthropod natural enemies: A database summary. Agriculture, Ecosystems and Environment. 1988;**21**(3-4):191-216

[5] Desneux N, Decourtype A, Delpuech JM. The sublethal effects of pesticides on beneficial arthropods. Annual Review of Entomology. 2007;**52**:81-106

[6] Zappala L, Bernardo U, Biondi A, Cocco A, Deliperi S, Delrio G, et al. Recruitment of native parasitoids by the exotic pest *Tuta absoluta* in southern Italy. Bulletin of Insectology. 2012;**65**:51-61

[7] Shirazi J, Morsali H, Rezaei M. A review on the side effects of chemical pesticides on the pests' natural enemies. (In Persian with English abstract). In: IRIPP, PPO, editor. Proceeding of Half a Century Pesticides Usage in Iran. 2009. pp. 103-132

[8] Tian HX, Lin NQ. The genus Trichogrammatoidea Girault from Hainan, with description of a new species (hymenoptera, Trichogrammatidae). Acta Zootaxonomica Sinica. 2009;**34**:360-366

[9] Sidi MB, Islam MT, Ibrahim Y, Dzolkhifli O. Effect of azadirachtin and rotenone on *Trichogramma papilionis* (Hym: Trichogrammatidae). Journal of Food Agriculture and Environmental. 2013;**2**:1509-1513

[10] Stark JD, Banks JE. Population-level effects of pesticides and other toxicants on arthropods. Annual Review of Entomology. 2003;**48**(505-519):30

[11] IOBC-WPRS Bulletin. Working group standard methods to test the side-effects of pesticides on natural enemies of insects and mites developed by the IOBC/WPRS Working Group 'Pesticides and Beneficial Organisms. Bulletin OEPP/EPPO Bulletin (Western Palearctic regional section). 2008;**15**(2):214-255

[12] Overmeer WPJ, Vanzon AG. A standardized method for testing the side effects of pesticides on the predacious mite *Amblyseius potentillae*. Entomophaga. 1982;**27**(4):357-363

[13] Sterk G, Hassan SA, Baillod M, Bakker F, Bigler F, Blümel S, et al. Results of the seventh joint pesticide testing programme carried out by the IOBC, WPRS-working group 'pesticides and beneficial organisms'. BioControl. 1999;**44**(1):99-117

[14] Al-Rajab AJ, Alhababy AM, Alfaifi T. Persistence of imidacloprid, acetamiprid and methomyl in qat leaves. Hellenic Plant Protection Journal. 2016;**9**(5):51-59

[15] Chailleux A, Biondi A, Han P, Tabone E, Desneux N. Suitability of the pest-plant system *Tuta absoluta*

(Lepidoptera: Gelechiidae) tomato for *Trichogramma* (hymenoptera: Trichogrammatidae) parasitoids and insights for biological control. Journal of Economic Entomology. 2013;**106**(6):2310-2321

[16] Khan HDT, Chailleux A, Tiradon M, Desneux N, Colombel E, Tabone E. Using new egg parasitoids (*Trichogramma spp*) to improve integrated management against *Tuta absoluta*. EPPO Bulletin. 2012;**42**(2):249-254

[17] Uma S, Jacob S, Lyla KR. Acute contact toxicity of selected conventional and novel insecticides to *Trichogramma japonicum* Ashmead (hymenoptera: Trichogrammatidae). Journal of Biopesticides. India. 2014;7:133-136

[18] Desneux N, Wajnberg E, Wychhuys AG, Burgio G, Arpaia S, Naryaez-Vasquez C, et al. Biological invasion of European tomato crops by *Tuta absoluta*: Ecology, geographic expansion and prospects for biological control. Journal of Pest Science. 2010;**83**:197-215

[19] Rui L, Satta A, Floris I. Effects of a azadirachtin-based formulation on the nontarget muscoid fly parasitoid *Muscidifurax raptor* (hymenoptera: Pteromalidae). Biological Control. 2008;**47**:66-70

[20] Ruiz L, Flores S, Cancino J, Arredondo J, Valle J, Diaz-Fleischer F, et al. Acute and population level toxicity of imidacloprid and fenpyroxymate on an important egg parasitoid *Trichogramma cacoeciae*. Ecotoxicology. 2011;**20**(6):1476-1484

[21] Shi ZH, Guo SJ, Lin WC, Liu SS. Evaluation of selective toxicity of five pesticides against *Plutella xylostella* (Lep: Plutellidae) and their side-effects against *Cotesia plutellae* (Hym: Braconidae) and *Oomyzus sokolowskii* (hymenoptera: Eulophidae). Pest Management Science. 2004;**60**:1213-1219

[22] Saber M, Hejazi MJ, Hassan SA. Effects of azadirachtin/Neemazal on different stages and adult life table parameters of *Trichogramma cacoeciae* (hymenoptera: Trichogrammatidae). Journal of Economic Entomology. 2004;**97**:905-910

[23] Costa MA, Muscardini VF, Gontijo PDC, Carvalho GA, Oliveira RLD, Oliveira HND. Sub lethal and transgenerational effects of insecticides in developing *Trichogramma galloi*. Ecotoxicology. 2014;**23**(8):1399-1408

[24] Souza JR, Carvalho GA, Moura AP, Couto MHG, Maia JB. Toxicity of some insecticides used in maize crop on *Trichogramma pretiosum* immature stages, Chilean. Journal Agricultural Research. 2014;**74**:234-239

[25] Poorjavad N, Goldansaz SH, Dadpour H, Khajeali J. Effect of essential oil on some biological and behavioral traits of *Trichogramma embryophagum* and *T. Evanescens*. BioControl. 2014;**59**:403-413

[26] Mahankuda B, Sawai HR, Gawande RW, Neharkar PS, Nagdeote VG. Effect of insecticide residues on the adult survival rate of *Trichogramma chilonis* under laboratory condition. Journal of Entomology and Zoology Studies. 2019;7(2):1349-1351

[27] Kumar A, Singh NN, Mishra VK. Toxicity of insecticides to egg parasitoid, *Trichogramma chilonis* Ishii under laboratory and semi-field conditions. Annals of Plant Protection Sciences. 2019;**27**(1):24-27

[28] Deshmukh YV, Undirwade DB, Dadmal SM. Effect of some newer

insecticides on parasitisation by *Trichogramma* species under laboratory condition. Journal of Entomology and Zoology Studies. 2018;**6**(3):228-231

[29] Mohapatra B, Shinde CU. Contact toxicity of different insecticides against egg parasitoid, *Trichogramma japonicum* Ashmead under laboratory condition. Journal of Entomology and Zoology Studies. 2021;**9**(1):134-139

[30] Carmo ED, Bueno A, Bueno RCOF. Pesticide selectivity for the insect egg parasitoid *Telenomus remus*. BioControl. 2010;**55**:455-464

[31] Carvalho GA, Goday MS, Parreira DS, Rezende DT. Effect of chemical insecticides used in tomato crops on immature *Trichogramma pretiosum* (hymenoptera: Trichogrammatidae). Revista Colombiana de Entomología. 2010;**36**(1):10-15

[32] Fontes J, Roja IS, Tavares J, Oliveira L. Lethal and sublethal effects of various pesticides on *Trichogramma achaeae* (hymenoptera: Trichogrammatidae). Journal of Economic Entomology. 2018;**111**:1-8

[33] Nekunam A, Afshari A, Nadimi A. Investigation of persistency and toxicity of compound insecticides on cotton bushes in order protection against *Trichogramma brassicae* Bezd. Iranian Cotton Research Journal. 2020;**8**(2):107-126

[34] Afshari A, Gorzalddin M, Mottaki E. Side effects of Indoxacarb and Lufenuron on *Trichogramma achaeae* (hymenoptera: Trichogrammatidae) in laboratory condition. Plant Protection. 2014;**37**(3):61-79

[35] Hussain D, Ali H, Qasim M, Khan J. Insecticidal susceptibility and effectiveness of *Trichogramma chilonis* as parasitoids of tomato fruit borer, *Helicoverpa armigera*, Pakistan. Journal. Zoology. 2015;**47**(5):1427-1432

[36] Khan MA, Ruberson JR. Lethal effects of selected novel pesticides on immature stages of *Trichogramma pretiosum* (hymenoptera: Trichogrammatidae). Pest Management Science. 2017;**73**:2465-2472

[37] Hewa-Kapuge S, Dougall SM, Hoffmann A. Effects of methoxyfenozide, indoxacarb, and other insecticides on the beneficial egg parasitoid *Trichogramma brassicae* (hymenoptera: Trichogrammatidae) under laboratory and field conditions. Journal of Economic Entomology. 2003;**96**(4):1083-1090

[38] Mollanorouzi F, Hesami S, Ghaibi M, Hoseini M, Fallahzadeh M. The lethal effect of indoxacarb and thiodicarb insecticides on the parasitoid wasp *Trichogramma brassicae*. Journal of Entomological Research. 2018;**11**(1):65-77

[39] Vaez N, Iranipour SH, Hejazi MJ. Effect of *bacillus thuringiensis* on *Trichogramma brassicae* life history parameters. Applied Research in Plant Protection. Iran. 2020;**9**(3):29-45

[40] Tabebordbar F, Shishehbor P, Ziaee M, Sohrabi F. Effects of field recommended concentrations of three different insecticides on life table parameters of the parasitoid *Trichogramma evanescens* (Hym.: Trichogrammatidae) under laboratory conditions. Plant Pest Research. 2020;**9**(4):11-23

[41] Shirazi J, Bahrami A, Rezapanah M. Side effects of five pesticides on *Trichogramma brassicae* Bezdenko (hymenoptera:

Trichogrammatidae). Journal of Agriculture and Forestry. 2018;**66**(1):57-68

[42] Ashtari S, Sabahi Q, Talebi Jahromi K. Survey of Parasitismic effect of two species of *Trichogramma* on eggs of *Tuta absoluta* under effect of pesticides. Journal of Vegetables. 2020;**4**(7):1-11

[43] Ashtari S. Efficacy of spinosad, imidacloprid, indoxacarb and chlorantraniliprole on *Tuta absoluta* and two species of parasitoid wasps. Journal of Applied Research in Plant Protection. 2022a;**11**(2):79-90

[44] Parsaeyan E, Safavi SA, Saber M, Poorjavad N. Effects of emamectin benzoate and cypermetrin on the demography of *Trichogramma brassicae*. Crop Protection. 2018;**110**:269-274

[45] Ashtari S, Sabahi Q, Talebi Jahromi K. Comparison of the insecticidal effect of several compounds on biological parameters of *Trichogramma brassicae*. Pesticides in Plant Protection Sciences. 2019;**8**(1):47-61

[46] Shafaghi F, Fourouzan M, Mahdavi V, Attaran MR. Effect of chlorantraniliprole, Flubendiamide, and Indoxacarb against tomato leaf miner, and side effects of chlorantraniliprole on the parasitoid wasp, *Trichogramma brassicae*. Plant Pest Research. 2023;**13**(1):39-49

[47] Parsaeyan E, Safavi SA, Saber M, Poorjavad N. Effects of emamectin benzoate and cypermethrin on the demography of *Trichogramma brassicae* Bezdenko. Crop Protection. 2017;**110**:269-274. DOI: 10.1016/j.cropro.2017.03.026, 1-6

[48] Saber M, Ghorbani M, Vaez N, Armak A. Effects of diazinon and fipronil on functional response of *Trichogramma brassicae* Bezdenko (Hym.; Trichogrammatidae) in the laboratory conditions. Journal of Crop Protection. 2020;**9**(2):275-283

[49] Bhargavi M. Effect of different insecticides on adult emergence of *Trichogramma chilonis* (Ishii). International Journal of Plant Protection. 2017;**10**(1):79-82

[50] Ashtari S. Survey of persistency of spinosad, imidacloprid and indoxacarb on *Trichogramma brassicae* and *T. evanescens*. Journal of Plant Protection. 2018;**33**(1):27-34

[51] Ashtari S, Sabahi Q, Talebi Jahromi K. Evaluation of toxicity of some biocompatible insecticides on *Trichogramma brassicae* and *T. evanescens* under laboratory and semi-field conditions. Journal of Crop Protection. 2018;**7**(4):459-469

[52] Ashtari S. Toxicity of tetraniliprole, chlorantraniliprole, lufenuron and thiocyclam insecticides on *Trichogramma brassicae* Bezdenko and *T. Evanescens* Westwood (hymenoptera: Trichogrammatidae) under laboratory and semi-field conditions. Plant Protection (Scientific Journal of Agriculture). 2022b;**45**(3):91-103. DOI: 10.22055/ppr.2022.17829

[53] Alsaedi G, Ashouri A, Talaei Hassanloui R. Evaluation of *bacillus thuringiensis* to control *Tuta absoluta* (Meyrick) (Lepidoptera: Gelechiidae) under laboratory conditions. Agricultural Sciences. 2017;**8**:591-599

[54] Nozad-Bonab Z, Hejazi MJ, Iranipour SH, Arzanlou M, Biondai A. Lethal and sublethal effects of synthetic and bio-insecticides on *Trichogramma brassicae* parasitizing *Tuta absoluta*. PLoS One. 2021;**16**(7):e0243334. DOI: 10.1101/2020.11.20.391003

[55] Jafari M, Rafieidastjardi H, Nouriqanbalani Q, Fathi AA, Hasanpour M, Ebadallahi A. Smoke toxicity of three plant essential oils on *Phthorimaea operculella* and its effect on the percentage of *Trichogramma brassicae* parasitism. Plant Pest Research. 2021;**11**(1):71-85

[56] Rafiqi Z, Sharifi M, Javar S, Mottaki A. Investigation the lethal effects of consult and Confidor insecticides on *Trichogramma brassicae* parasitoid wasp eggs in laboratory conditions. In: National Conference on New Researches in Agricultural Engineering, Environment and Natural Resources in Iran. 2016. pp. 1-8

[57] Jafarinodooshan A, Shamszadeh M, Sami MA, Khalili M. Effect of the stinking Assa extract on parasitism and mortality rate of *Trichogramma* wasp as carob moth egg parasitoids (*Trichogramma brasicae*) *in vitro*. National Congress of Pomegranate in Iran 2012. pp. 605-609

[58] Khorrami F, Soleimanzadeh A, Forouzan M, Nouri H, Alikhani M. Lethal and deterrent efficacy of three plants extracts on the potato tuber moth, *Phthorimaea operculella* Zeller and influence on percent parasitism of *Trichogramma brassicae*. Biological in Plant Protection. 2019;**6**:77-87. (In Farsi)

[59] Alimirzanejad H, Vahedi H, Sadeghi A, Dadpourmoghanloo H. Investigation of side effects of four fungicide on *Trichogramma brassicae* [thesis]. Kermanshah, Iran: Razi University of Kermanshah; 2012. 54 p

[60] Costa MA, Farias ES, Passos LC, Carvalho VC, Carvalho GA. Side effects of insecticides applied to cotton on adult *Trichogramma pretiosum* by three exposure routes. Pest Management Society. 2022;**78**:1895-1902

[61] Parreira DS, Cruz RA, Zanuncio JC, Lemes GC, Rolim GS, Barbosa LR, et al. Essential oils cause detrimental effects on biological parameters of *Trichogramma galloi* immatures. Journal of Pest Science. 2018;**91**:887-895

[62] Alano DM, Araujo ES, Mirás-Avalos JM, Pimentel IC, Zawadneak MAC. Short communication: Sublethal effects of insecticides used in strawberry on *Trichogramma pretiosum* (hymenoptera: Trichogrammatidae). Spanish Journal of Agricultural Research. 2021;**19**(1):e10SC01. DOI: 10.5424/sjar/2021191-17235

[63] Sabahi Q. Pesticide Application Technology. University of Tehran Press; 2019. 294 p

Chapter 7

WASPAS and in the Study and Geospatial Analysis of the Abundance of Hymenopteran Species in Urban Areas of Caribbean Zone

Hiovanis Castillo Pantoja, Yaimy Marrero Pérez and Ricardo Lorenzo Avila Rondon

Abstract

The scientific article presented here focuses on the use of WASPAS as a multi-criteria tool and geospatial analysis to study the abundance of hymenopteran species in urban areas of the Caribbean. WASPAS is a tool that is based on the collection of insect observation data, and its subsequent analysis using statistical techniques allows to reach the ordered ranking of criteria so that the results obtained allow to know the abundance of different insect species in urban areas and their evolution over time, considering environmental and risk criteria. It is possible to identify the most common insect species in these areas and their relationship with environmental factors such as temperature and humidity, using geospatial tools. The result obtained highlights the importance of this tool for biodiversity monitoring in urban areas and for decision-making in the management of green spaces and urban planning.

Keywords: WASPAS, geospatial analysis, urban areas, biodiversity, green spaces

1. Introduction

There are several methods to carry out studies of hymenopteran species. Among the studies of the Caribbean and South American areas, we focus on Ref. [1], which through the study of insects in plantations evaluates the potential of using images to detect ant networks in *Tectona grandis* plantations in Brazil using partial least squares discriminant analysis (PLS-DA) models. Similarly, Ref. [2] analyzes the ecological importance of the ecosystems provided by Bumble bees (Hymenoptera: Apidae, *Bombus latreille*) and the changes that appear from anthropological behavior, as influenced by the actions of insecticides and others that influence the way of life of bees. De Lima Santos et al. [3] relies on the use of datasets to carry out their study for the comparison of factors of the diversity, biology, biogeography, and evolution of

Anthidium Fabricius in the Western Hemisphere. The dataset in the research of the 92 species of Anthidium with their main characteristics is grouped into the variables sex, temporal and geospatial details, coordinates, data collector, host plants, associated organisms, name of identifier, historic identification, historic identifier, taxonomic value. Faizan et al. [4] makes its contribution on a similar topic, only that it focuses on the populations of Bumble bees and the importance they represent in the pollination of plants, for the economies of agricultural productions. In research on the analysis of bees and honey production. Gratton et al. [5] combines microscopic analysis of pollen and patterns described in spatial representation on soils to help determine the interactions of various factors in habitat types in the regions studied Ref. [6] develops a whole exposition in relation to the importance of conservation of bee species (Hymenoptera, Apidae, and Meliponini) and their pollination work in tropical lands and how deforestation actions have negatively impacted the sizes of bee communities. For this purpose, it uses spatial scales as a tool, combining several criteria such as forest area, forest edge, and connectivity of forest patches. In this context, it is shown that research using multi-criteria analysis, specifically MCDM techniques and spatial representation of data, is a very useful tool. In the exhibition of Ref. [7] studies the phenomenon of erosion by water; they work on the parameters obtained, using sophometry. Kayal et al. [8] complements this topic by analyzing semiarid zones. In another area, we find the effects of floods in which Ref. [9] obtains thematic maps of flood zones with high, medium, and moderate degrees and combines climate and physical-graphic criteria, which allows cities to plan evacuation plans for these contingencies. In this context, it is shown that research using multi-criteria analysis, specifically MCDM techniques and spatial representation of data, is a very useful tool. Such is the case in [10] studies the phenomenon of erosion by water; he combines for them the parameters obtained from the use of sophometry. Lin et al. [11] complements this topic by analyzing semi-arid zones. In another area we find the effects of floods in which [12] obtains thematic maps of flood zones with high, medium and moderate degree and combines climate and physical-graphic criteria, which allows cities to plan evacuation plans for these contingencies. In relation to studies in urban areas with the combination of MCDM-GIS tools, we start from Ref. [13] that incorporates in the research the analysis of uncertainty in urban areas of India in relation to the phenomena of alluvium formation, allowing the checking of water levels and thus allowing better decision-making in the face of these hydrological phenomena. Dagdeviren et al. [14] similarly takes this approach but in terms of water treatment and its relationship with the high volumes of population, analyzing the best place for its location, working on the basis of 9 criteria that allowed it to arrive at 5 ways of evaluating the locations of the plants by a degree of hierarchy. Kamal and Al-Subhi [15] focuses on groundwater and its rise in the use by populations in areas such as domestic uses, irrigation, sectors such as agriculture, and so on. Patil and Gupta [16] proposes, based on the physical-GIS-MCDM analysis, to obtain the hierarchy of hydrological and socioeconomic criteria, among others, that will allow better management based on the phenomena that provide temporal patterns, with the frequency and duration in which they occur in order to reduce material losses.

Rana and Moniruzzaman and Yu et al. [17, 18] classify the lands susceptible to urban floods from their GIS representations, which allows with the spatial representation to identify the main risks. Peng et al. [19] comes to enhance the structures to use rainfall in crops, reaching the locations with greater potential in urban districts.

As can be seen, the relationship established between MCDM and GIS is a very useful tool, current and with a solution to problems by establishing a priority order.

Among the best methods for the author is the use of Weighted Aggregated Sum Product Assessment (WASPAS) [16], highlights in its observation the climatic conditions in different parts of the world affecting species, using this WASPAS method evaluates the potential of viable criteria to be taken into account for the construction of Rainwater Harvesting (RWH) and its integration to geographic information systems (GIS).

This evaluation reaffirms that by using MCDM-GIS, it is possible to establish, based on quantitative criteria, the hierarchy of the areas with presence of hymenopteran species in urban areas of an island in the Caribbean zone. In this way, control activities and conservation management of insect species can be organized.

2. WASPAS method

The WASPAS method is a unique combination of two well-known MCDM approaches, that is, weighted sum model (WSM) and weighted product model (WPM). Its application first requires the development of a decision/evaluation matrix, $X = [x_{ij}]$ m × n where x_{ij} is the performance of i_{th} alternative with respect to j_{th} criterion, m is the number of alternatives, and n is the number of criteria. The methodology to be followed to apply the method is shown in **Figure 1** [1–6].

Variances of estimates of alternatives in WASPAS method depend of the variances of WSM and WPM approaches as well as on the value of λ. It may be worthwhile to compute the optimal values of λ and assure the maximum accuracy of estimation. It may also be important to study the effects of optimal λ values on the final ranking of the alternatives.

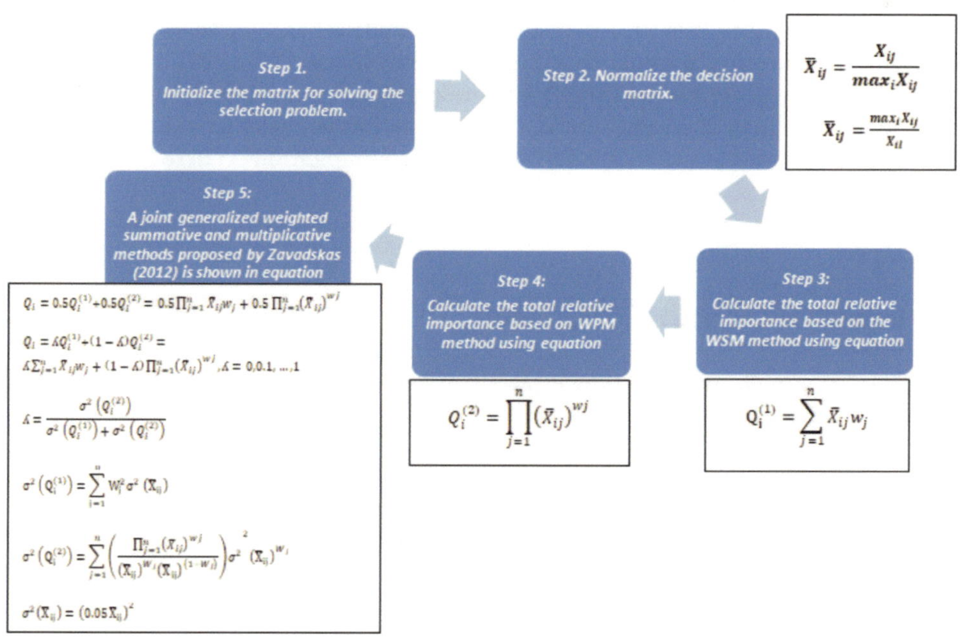

Figure 1.
The structure of the steps of the WASPAS method.

3. Evaluating weights by using AHP method

3.1 Determination of the weights of the criteria

The index of weights for the MCDM method is performed by the analytical hierarchical process (AHP) method. This method proposed by Saaty is a widely used tool for modeling unstructured problems, synthesizing subjective and objective information found in many disciplines, such as: politics, economics, biology, sociology, management science, and biology and related sciences.

The steps of the AHP method are structured as follows:

1. Develop a model for the business

2. Derive priorities (weights) for the criteria

3. Consistency check (weight assigned correct or not?)

4. Derive overall priorities (Model Synthesis) and final decision.

4. Case study

The study focused on the city of Holguín, belonging to the eastern city with the same name; it is located at coordinates 20°53′14″N 76°15′47″W/20.887222222222, −76.263055555556. The city has an urban extension of 9210 km^2, a population of 1,032,670 inhabitants, with a density of 111 hab/km^2. The rural areas extend between 5 and 9 km from the urban area. According to studies, there is a level of insects in different urban and rural areas.

The systematic sample in the urban areas of the study is based on the collection of data on the presence and abundance of hymenopteran species. We worked on the sample of 42 observations presented in **Table 1**. For the analysis of the study, the evaluation of the clusters with the MCDM algorithm used the quantitative variables of temperature, humidity, number of insects, insect density, and the precipitation occurred.

5. Analysis

5.1 Determination of criteria weights

To run the model, by stating the established criteria and using the equations of the numbers, the weights of the criteria are evaluated and the dimensional deviation is taken into account.

The scale table pairwise comparison show in the **Table 2**. Shows the initial decision matrix, in which the criteria are evaluated, in the **Table 3**.

Figure 2 shows the results of lateral deviation as the criterion with the highest weight: $W_{temperature} = 0.05$, $W_{humedity} = 0.28$ y $W_{number\ of}$ Insects = 0.24, $W_{densidity} = 0.30$, $W_{precipitation} = 0.12$. According to the hierarchy, insects' density for square meters in the one belonging to Mayr I weigh for the experts' decision.

In this, it is determined that the criterion of population density is estimated with a weight of 30%, given the need to know the level of presence of insects. Next, the

No	Month	Area	Temperature (°C)	Time of day	Humidity (%)	Number of insects	Location	Insect family	Insect family density (insects/m^2)	Precipitation (mm)
1	January	Town center	20	Morning	60	50	Residencial	Paratrechina longicornis	5	10
2	January	Park	26	Afternoon	55	45	Park	Atta insulares	3	5
3	February	Trading area	23	Night	50	21	Residencial	Atta insulares	7	2
4	February	Industrial zone	24	Afternoon	54	22	Fabril	Braconidae	9	0
5	February	Area surrounding a commercial port	22	Night	50	15	Muelle 1	Braconidae	5	1
6	March	Rural residence	25	Morning	52	21	Valle	Braconidae	7	0
7	March	Airport	23	Afternoon	54	15	Zona AB	Apis cerana	3	6
8	March	Highway	28	Night	53	15	Km 102	Helictidae	3	3
9	March	Town center	24	Morning	59	3	Centro Sur	Dorymyrmex insanus	7	0
10	March	Park	25	Afternoon	61	11	La Quinta	Atta insulares	8	4
11	April	Trading area	26	Night	60	12	El Jardín	Paratrechina longicornis	6	3
12	April	Area Hospital	28	Morning	56	14	Bloque Este	Dorymyrmex insanus	7	0
13	April	Industrial zone	27	Afternoon	55	5	Industria Construcciones	Apis cerana	5	5
14	April	Area surrounding A commercial port	25	Night	54	12	Muelle 5	Helictidae	7	3
15	April	Rural residence	26	Morning	50	6	Finca Mayabe	Braconidae	5	2
16	May	Airport	29	Afternoon	55	9	Cabeza de Pista	Apis cerana	6	5

No	Month	Area	Temperature (°C)	Time of day	Humidity (%)	Number of insects	Location	Insect family	Insect family density (insects/m^2)	Precipitation (mm)
17	May	Highway	27	Night	57	11	Trebol Norte	Dorymyrmex insanus	9	2
18	May	Town center	29	Morning	50	8	Centro Sur	Atta insulares	7	0
19	May	Park	28	Afternoon	52	12	Vista Alegre	Paratrechina longicornis	4	3
20	May	Trading area	27	Night	54	15	Mercado Sur	Dorymyrmex insanus	5	5
21	May	Area Hospital	26	Morning	51	6	Área de Lavandería	Paratrechina longicornis	8	1
22	June	Industrial zone	26	Afternoon	53	9	Fabril KTP	Braconidae	5	3
23	June	Area surrounding A commercial port	25	Night	54	12	zona de enlace	Apis cerana	6	2
24	June	Rural residence	22	Morning	55	7	Corrales del Norte	Helictidae	8	2
25	June	Airport	26	Afternoon	56	15	Área del Canal	Apis cerana	5	3
26	July	Highway	23	Night	52	14	Sección Trebol	Dorymyrmex insanus	7	4
27	July	Town center	23	Morning	52	9	Central Tecnológico	Atta insulares	8	3
28	July	Park	27	Afternoon	54	22	Plaquita	Paratrechina longicornis	11	2

Table 1.
Sample of observations taken for the study.

01-Sep	01-Jul	01-May	01-Mar	1	3	5	7	9
Extremely	Very strongly	Strongly	Moderately	Equally	Moderately	Strongly	Very strongly	Extremely

Table 2.
Rating for pairwise comparison according to Saaty (2008).

	Temperature (°C)	Humidity (%)	Number of insects	Density (insects/m^2)	Precipitation (mm)
Temperature (°C)	1	0.1	0.2	9	5
Humidity (%)	7	1	9	9	7
Number of Insects	5	7	1	7	7
Density (insects/m^2)	9	9	9	1	8
Precipitation (mm)	7	7	0.2	7	1

Table 3.
Initial decision matrix.

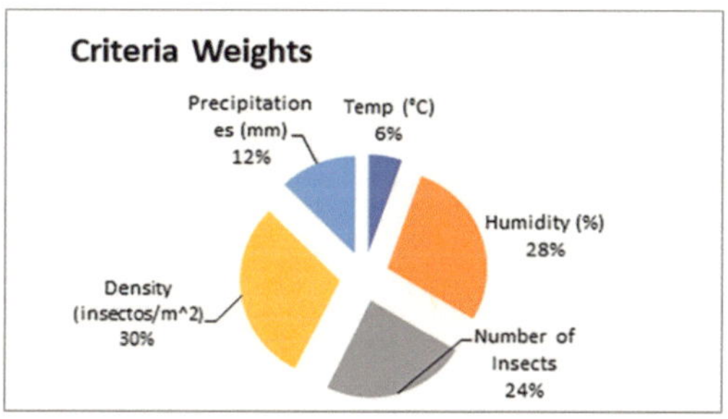

Figure 2.
Result of the weights.

humidity variable is determined in second place because it is a key ecological factor that influences the survival and development of insect populations, in addition to maintaining its observation that allows us to predict the population dynamics of the species under study. Subsequently, we find the number of insects that influence the density and close consecutively the precipitation and temperature that are taken into account from the specific conditions carried out in this study.

6. WASPAS with a single value set (WASPAS)

The solution of the WASPAS method begins with the normalization of the data in **Table 4**, taking the values of the results of the experimental work with the initial parameters.

When analyzing the 47 alternatives of the study, according to the defined weights of the study variables, the WASPAS method for an optimal solution with $\lambda = 0.5$ establishes a hierarchical result of the alternatives 41-47-46-1-2-4-9-33-10-34-35-34-35-24-37-31-28-45-16-39-43-22-29-42-27-25-38-30-30-32-11-11-17-18-17-3-5-23-21-15-40-20-20-13-44-19-12-12-26-6-8-36-14-7, which are listed in **Table 5**.

No. alternativa	Matriz de decision normalizada				
1	1	0.8333	1	0.4545	0.1
2	0.7692	1.1	0.9	0.2727	0.2
3	0.8696	1	0.42	0.6364	0.5
4	0.8333	1.08	0.44	0.8182	1
5	0.9091	1	0.3	0.4545	1
6	0.8	1.04	0.42	0.6364	1
7	0.8696	1.08	0.3	0.2727	0.1667
8	0.7143	1.06	0.3	0.2727	0.3333
9	0.8333	1.18	0.06	0.6364	1
10	0.8	1.22	0.22	0.7273	0.25
11	0.7692	1.2	0.24	0.5455	0.3333
12	0.7143	1.12	0.28	0.6364	1
13	0.7407	1.1	0.1	0.4545	0.2
14	0.8	1.08	0.24	0.6364	0.3333
15	0.7692	1	0.12	0.4545	0.5
16	0.6897	1.1	0.18	0.5455	0.2
17	0.7407	1.14	0.22	0.8182	0.5
18	0.6897	1	0.16	0.6364	1
19	0.7143	1.04	0.24	0.3636	0.3333
20	0.7407	1.08	0.3	0.4545	0.2
21	0.7692	1.02	0.12	0.7273	1
22	0.7692	1.06	0.18	0.4545	0.3333
23	0.8	1.08	0.24	0.5455	0.5
24	0.9091	1.1	0.14	0.7273	0.5
25	0.7692	1.12	0.3	0.4545	0.3333
26	0.8696	1.04	0.28	0.6364	0.25
27	0.8696	1.04	0.18	0.7273	0.3333
28	0.7407	1.08	0.44	1	0.5
29	0.8	1.06	0.42	0.6364	0.25
30	0.8696	1.18	0.22	0.7273	0.5
31	0.6897	1.22	0.18	0.7273	0.5
32	0.7692	1.2	0.26	0.6364	0.5
33	0.8	1.02	0.2	0.4545	1
34	0.6897	1.06	0.28	0.6364	0.5
35	0.7692	1.08	0.32	0.7273	0.3333
36	0.8333	1.1	0.12	0.2727	0.5
37	0.7692	1.04	0.26	0.7273	0.25
38	0.8696	1.04	0.3	0.6364	0.3333
39	0.9524	1.08	0.18	0.8182	0.3333
40	0.7692	1.06	0.34	0.7273	0.1667

Table 4.
FMS normalized data is created.

Eq. 4, Q2							Score
0.9901	1.0218	0.8231	0.941	1	0.7836	1.1744	1
0.9948	1	0.7516	0.7875	1	0.5888	0.9815	2
0.9799	1.0574	0.6658	0.908	0.9173	0.5746	0.9496	3
0.9879	1.0111	0.814	0.872	1	0.709	0.9019	4
0.9799	1.0574	0.6658	0.908	0.9173	0.5746	0.9496	3
0.9857	1.0525	0.7265	0.872	0.9173	0.6029	0.8579	5
0.9857	1.0525	0.7632	0.872	0.9173	0.6333	0.8497	6
0.9924	1.0271	0.7516	0.908	0.8721	0.6066	0.8474	7
0.9879	1.0056	0.7742	0.941	0.8183	0.5922	0.847	8
0.9924	1.0218	0.7516	0.6746	0.7999	0.4113	0.8369	9
0.9901	1.0475	0.5131	0.872	1	0.464	0.825	10

Table 5.
Optimum values per hierarchy for λ = 0.5.

Table 6 shows the data from the study carried out for the first 10 alternatives determined by the hierarchy of the WASPAS method, as a result of its multi-criteria analysis.

Based on the results obtained, the authors of the research present the following action plan as a proposal, structured in a table and providing a clear vision of the measures, objectives, actions, responsible parties, deadlines, and resources necessary to protect and conserve Hymenoptera species, starting with the 10 selected areas in **Table 7**.

7. Conclusions

The purpose of this study is to improve our knowledge for the present time, the decision-making in relation to treatment planning and control of the abundance of hymenopteran species in urban areas in Caribbean islands, as achieved with the use of MCDM tools and spatial location. The characteristics provided by the quantitative variables used and establishing their weights determined that the density of insects ($W_{density}$ = 0.30) is the one with the highest weight, by decisions of the experts, which responds to the perspective of knowing the distribution of the different populations for their ecological analysis and the vulnerabilities to which they are subjected in their habitat for the areas studied. Subsequently, the WASPAS method is used to combine the 5 quantitative criteria to arrive at the first 10 records and to propose an action plan with the projection of possible future scenarios considering the evolution of key factors such as insect resistance and climate change, as well as to evaluate different insect management options considering multiple criteria, anticipating their future consequences. This orients decision-making towards effective, sustainable alternatives adapted to changing conditions with the hymenopteran species specified in the study in the city of Holguin. This result constitutes a novelty for this type of study in particular and one that can be extended to other types of areas and with other species.

Conflict of interest

The authors declare that they have no conflicts of interest.

No	Month	Area	Temperature (°C)	Time of day	Humidity (%)	Number of insects	Location	Density (insects/m^2)	Insect family	Precipitation (mm)	Score
47	December	Airport	23	Night	55	15	Rampa 7	8	Braconidae	3	1
46	December	Rural residence	27	Afternoon	54	11	Pedernales	9	Braconidae	3	2
1	January	Town center	20	Morning	60	50	Residencial	5	Paratrechina longicornis	10	3
2	January	Park	26	Afternoon	55	45	Park	3	Atta insulares	5	4
4	February	Industrial zone	24	Afternoon	54	22	FApril	9	Braconidae	0	5
9	March	Town center	24	Morning	59	3	Centro Sur	7	Dorymyrmex insanus	0	6
33	August	Trading area	25	Morning	51	10	Mercado Crisol	5	Solenopsis geminata	0	7
10	March	Park	25	Afternoon	61	11	La Quinta	8	Atta insulares	4	8
34	August	Área Hospital	29	Afternoon	53	14	Lenin	7	Atta insulares	2	9
35	August	Industrial zone	26	Night	54	16	Empresa Eléctrica	8	Solenopsis geminata	3	10

Table 6.
Data for the 10 alternatives with priority of analysis.

Measure	Objective	Actions	Responsible	Timeframe	Resources
1. Habitat conservation	Protect and restore key habitats for Hymenoptera species	1. Identify areas with high diversity of Hymenoptera	Species Conservation	6 months	$100,000
		2. Implement programs to restore degraded habitats	Environmental organizations	1 year	$500,000
2. Education and awareness	Educate and raise public awareness of the importance of Hymenoptera species	1. Develop educational programs in schools and communities	Ministry of Education	9 months	$50,000
		2. Conduct public awareness campaigns Environmental organizations	Environmental organizations	1 year	$200,000
		3. Train volunteers to participate in monitoring and conservation activities	Conservation organizations	1 year	$30,000
3. Research and monitoring	Research and monitor Hymenoptera populations to inform decision-making	1. Conduct population surveys to assess the status of Hymenoptera species	University researchers	1 year	$150,000
		2. Establish monitoring programs to track the health and distribution of bee, wasp, and ant populations	Conservation organizations	2 years	$300,000
		3. Support scientific research on the ecology, behavior, and threats faced by Hymenoptera species	Scientific institutions	3 years	$500,000

Total Budget: $2,530,000 Total Term: 3 years.

Table 7.
The action plan.

Author details

Hiovanis Castillo Pantoja[1*], Yaimy Marrero Pérez[1] and Ricardo Lorenzo Avila Rondon[2]

1 Universidad de Holguín, Holguín, Cuba

2 Autonomous University of Coahuila, Saltillo, Mexico

*Address all correspondence to: hiovaniscp@uho.edu.cu

References

[1] Arabameri A, Tiefenbacher JP, Blaschke T, Pradhan B, Bui DT. Morphometric analysis for soil erosion susceptibility mapping using novel GIS-based ensemble model. Remote Sensing. 2020;**12**(5):874. DOI: 10.3390/rs12050874

[2] Badapalli PK, Nakkala AB, Kottala RB, Gugulothu S. Geo environmental green growth towards sustainable development in semi-arid regions using physicochemical and geospatial approaches. Environmental Science and Pollution Research. 2022;**1-18**. DOI: 10.1007/s11356-022-24588-z

[3] de Lima Santos IC, dos Santos A, Oumar Z, Soares MA, Silva JCC, Zanetti R, et al. Remote sensing to detect nests of the leaf-cutting ant atta sexdens (hymenoptera: Formicidae) in teak plantations. Remote Sensing. 2019;**11**(14):1641. DOI: 10.3390/rs11141641

[4] Faizan MO, Hudait M, Sengupta K, Chowdhury BR. Flood susceptibility mapping using MCDM-AHP approach and geospatial techniques—a study of Hyderabad District, Telangana, India. International Journal of Hydrology Science and Technology. 2023;**16**(2):204-221. DOI: 10.1504/ijhst.2023.132602

[5] Gratton EM, McNeil DJ, Grozinger CM, Hines HM. Local habitat type influences bumble bee pathogen loads and bee species distributions. Environmental Entomology. 2023;**52**(3):491-501. DOI: 10.1093/ee/nvad027

[6] Griswold T, Gonzalez VH, Ikerd H. AnthWest, occurrence records for wool carder bees of the genus *Anthidium* (hymenoptera, megachilidae, Anthidiini) in the Western hemisphere. ZooKeys. 2014;**408**(408):31-49. DOI: 10.3897/zookeys.408.5633

[7] Jesiya NP, Gopinath G. A fuzzy based MCDM–GIS framework to evaluate groundwater potential index for sustainable groundwater management—a case study in an urban-periurban ensemble, southern India. Groundwater for Sustainable Development. 2020;**11**:100466. DOI: 10.1016/j.gsd.2020.100466

[8] Kayal P, Majumder S, Chowdhury IR. Modeling the spatial pattern of potential groundwater zone using MCDM-AHP and geospatial technique in sub-tropical plain region: A case study of Islampur sub-division, West Bengal, India. Sustainable Water Resources Management. 2022;**8**(6):185. DOI: 10.1007/s40899-022-00759-1

[9] Kaykhosravi S, Khan UT, Jadidi MA. A simplified geospatial model to rank LID solutions for urban runoff management. Science of the Total Environment. 2022;**831**:154937. DOI: 10.1016/j.scitotenv.2022.154937

[10] Koch JB, Lozier J, Strange JP, Ikerd H, Griswold T, Cordes N, et al. USBombus, a database of contemporary survey data for North American bumble bees (hymenoptera, Apidae, Bombus) distributed in the United States. Biodiversity Data Journal. 2015;**3**(3):e6833. DOI: 10.3897/bdj.3.e6833

[11] Lin C-H, Suresh S, Matcham E, Monagan P, Curtis H, Richardson RT, et al. Soybean is a common nectar source for honey bees (hymenoptera: Apidae) in a Midwestern agricultural landscape. Journal of Economic Entomology. 2022;**115**(6):1846-1851. DOI: 10.1093/jee/toac140

[12] Majed S, Ghafour Z. Geospatial analysis model for locating optimum centralized wastewater treatment plant for Sulaimania city. Sulaimani Journal for Engineering Sciences. 2023;**9**(2):57-73. DOI: 10.17656/sjes.10156

[13] Mayes DM, Bhatta CP, Shi D, Brown JC, Smith DR. Body size influences stingless bee (hymenoptera: Apidae) communities across a range of deforestation levels in Rondônia, Brazil. Journal of Insect Science. 2019;**19**(2):23. DOI: 10.1093/jisesa/iez032

[14] Dagdeviren M, Yavuz S, Kılınç N. Weapon selection˘ using the AHP and TOPSIS methods under fuzzy environment. Expert Systems with Applications. 2009;**36**(4):8143-8151

[15] Kamal M, Al-Subhi AH. Application of the AHP in project management. International Journal of Project Management. 2001;**19**(1):19-27

[16] Patil D, Gupta R. GIS-based multi-criteria decision-making for ranking potential sites for centralized rainwater harvesting. Asian Journal of Civil Engineering. 2022;**24**(2):497-506. DOI: 10.1007/s42107-022-00514-z

[17] Rana MSP, Moniruzzaman. A combined GIS, remote sensing and MCDM approach to find potential location for rainwater harvesting structure in northwestern part of Bangladesh. Hydro Research. 2023;**6**:235-246

[18] Yu X, Guo S, Guo J, Huang X. Rank B2C e-commerce websites in e-alliance based on AHP and fuzzy TOPSIS. Expert Systems with Applications. 2011;**38**(4):3550-3557

[19] Peng Y, Kou G, Wang G, Wu W, Shi Y. Ensemble of software defect predictors: An AHP-based evaluation method. International Journal of Information Technology & Decision Making. 2011;**10**(1):187-206

Chapter 8

Alien Species: *Vespa Velutina Nigrithorax* (Hymenoptera: Vespidae) – Proliferation and Methods for Its Control

Omaira de la Hera, Maria Luz Alonso and Rosa Maria Alonso

Abstract

Vespa velutina nigrithorax has become an invasive species since its introduction in Europe, as it is a predator of native fruits and insects, mostly honeybees. In this chapter, the morphology of the *Vespa velutina* is described together with its life cycle and the morphology of the nest that this hornet builds. The proliferation of this invasive species in Europe and the economic, environmental and human health impact that it produces makes necessary the use of methods for its control. The methods for the inactivation of nest, trapping hornets and minimising the impact on apiaries are the established methods to fight against this species. Behavioural study of these hornets in captivity has been carried out to increase the knowledge on the ethology of *Vespa velutina.* For this purpose, two secondary nests and one embryonic nest were captured and kept under controlled environmental conditions for up to 13 weeks for the secondary nest and 6 weeks for the embryonic nest. Adaptation to captivity, defence against disturbance, colony evolution and hibernation were the different behaviours observed. The results of this research will allow us to obtain additional information on this species, which are crucial to develop effective control methods for this invasive species.

Keywords: *Vespa velutina nigrithorax*, life cycle, morphology, proliferation, control methods, captivity study

1. Introduction

Vespa velutina nigrithorax (Lepeletier, 1836) is a eusocial hymenopteran belonging to the genus *Vespa*, one of the four genera in the subfamily Vespinae [1–3]. This species, better known as the Asian or yellow-legged hornet, like most *Vespa* species, is native to the Asian continent, extending from Afghanistan to Eastern China, Indochina and the Indonesian Archipelago [1, 4, 5].

This type of insect is characterised by the most highly developed social behaviour, including some species of bees, hornets, wasps, ants and termites.

IntechOpen

As a eusocial insect, *Vespa velutina* spends most of its life cycle in large colonies of hundreds to thousands of individuals, organised in castes, with different assigned roles [6, 7].

The caste system of Asian hornet is divided into queen, workers and males. The queen is responsible for the laying of eggs, from which new individuals will hatch. The workers, who are females, take care of the other functions. They feed the colony by collecting proteins, sugars and water, which will be supplied by trophallaxis, from adults to larvae and *vice versa*. This way of feeding means eusocial insects are not able to survive alone. They are also responsible for building, cleaning and repairing the nest, as well as protecting it, acting as guardians at the nest's entrance. The males have as their only function to fertilise the new queens [7–9].

2. *Vespa velutina's* morphology

Vespa velutina is physically similar to *Vespa crabro* (Linnaeus, 1758), which is the native European hornet, although they are easy to differentiate. *Vespa velutina* is slightly smaller, measuring between 2.5 and 3 cm, while *Vespa crabro* can grow up to 3.4 cm. Their colours are completely different, which is the best feature to differentiate between the two species. *Vespa velutina* has an orange head and a black face, and the thorax is entirely velvety black. The abdomen is also blackish except for the fourth segment, which is orange, as well as a thin orange band between the first and second segment. The dark colour of its body continues up to the middle of its legs, which are yellow at the ends. Its wings are dark with brownish tones.

Vespa crabro has a reddish head and yellow face. Its thorax is a combination of reddish-brown and its abdomen has the first two black segments separated by a yellow band, while the rest are yellow. Its legs are completely brown, the same colour as the wings (**Figure 1**).

Other characteristic of both hornet species is the shape of the end of the abdomen to identify sex differences. In males, it is flattened because they lack a sting, while in females it is conical, where the sting is stored (**Figure 2**).

Figure 1.
Differences between Vespa velutina (left) and Vespa crabro (right). Source: Image modified from http://anti-frelon-d-asie-jp33.over-blog.com (accessed on 9 January 2024) and from [10].

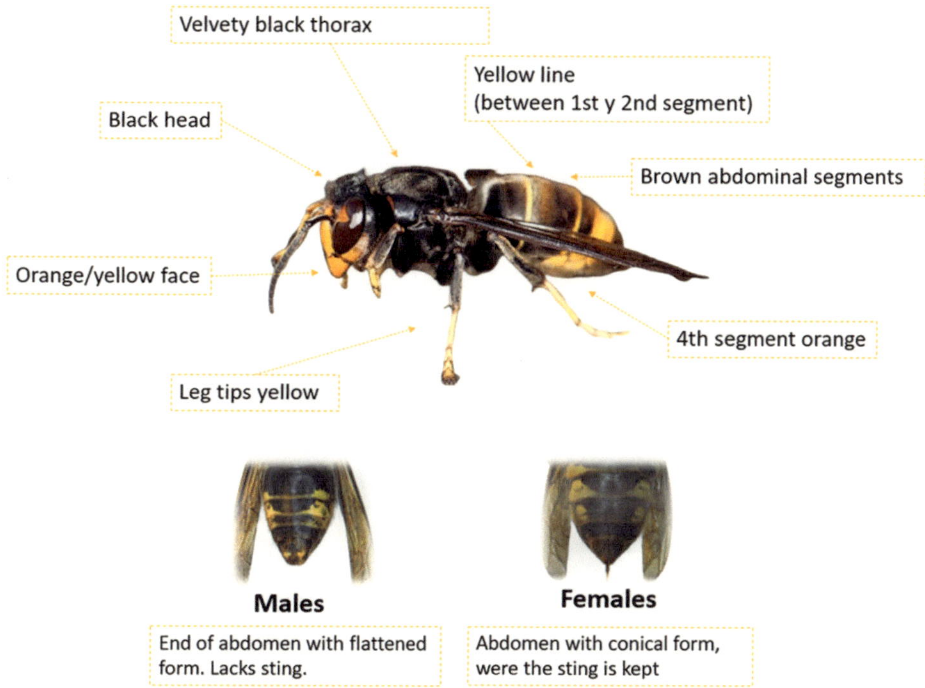

Figure 2.
Physical characteristics of Vespa velutina and differentiation between males and females. Source: Images modified from J. Urkiola and M. Lopez of Apicultura Ibérica journal.

3. Life cycle

The life cycle of *Vespa velutina* (**Figure 3**) is similar to the life cycles of other species of *Vespa* [12]. It starts when the queen emerges from its lethargy, who was overwintering, builds the embryo nest in a protected site and lays the first eggs. This time is the most vulnerable for the queen because she has to forage for itself and hunt to feed the first larvae alone until the first workers emerge, as well as collecting material for building the nest [7, 8, 13].

The larvae are fed on various insects and arachnids, which provide protein, while adult individuals, on the other hand, are unable to digest protein, and their diet is based on sugary liquids (nectar from flowers, blossoms or fruit) [14, 15]. In addition, after eating, the larvae secrete a substance rich in amino acids, which serve as food for the workers and the queen in case of need [16].

Between late spring and early summer, the colony increases, causing a lack of space in the primary nest, which leads to the construction of a secondary nest. They usually abandon the embryonic nest and build a new one in the upper part of leafy trees. In midsummer, due to the large number of larvae in the nest, this is the phase that gives rise to a great deal of insect predation, especially in apiaries of bees. From the construction of the secondary nest until autumn, hundreds to thousands of hornets may emerge [17].

With the arrival of autumn starts the reproduction season, the gynes (future queens) and males, from another colony, mate outside the nest. This is to avoid the inbreeding of the species. With the onset of winter, due to lack of food, comes the

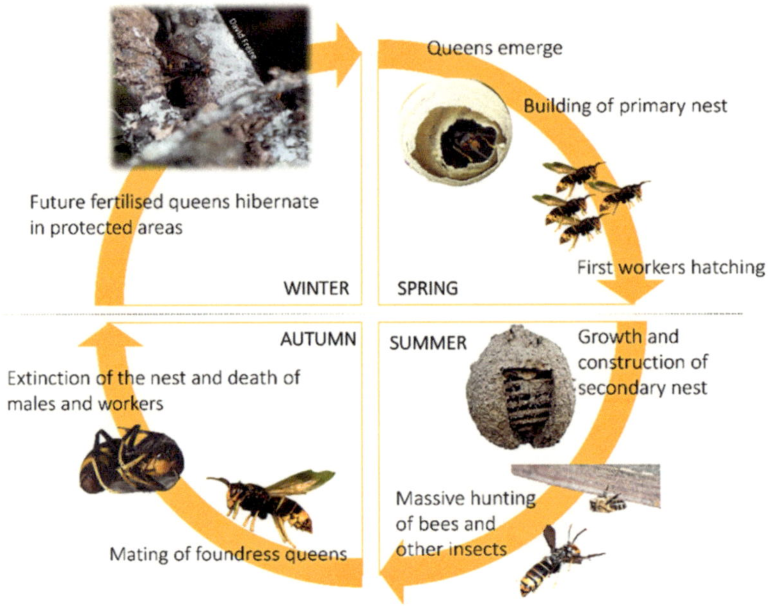

Figure 3.
Life cycle of Vespa velutina. Figure slightly modified from [11].

extinction of the nest with the death of the queen, workers and male, while gynes mated and unmated search a place to overwinter until next spring, when the new cycle starts [8–10, 13].

4. Nests' morphology

The nests of *Vespa velutina* (**Figure 4**), like other species of wasps and hornets, are mainly made of bark from different trees. The bark is shredded by the hornets with their mandibles and mixed with saliva to form a paste with which they shape the nest [18, 19].

Figure 4.
Primary nest (left) and secondary nest (right) of the invasive species Vespa velutina. Source: Image modified from J.M. Perez de Ana.

As indicated in the section on the life cycle, *Vespa velutina* generally builds two nests during its lifetime. Although both have similarities in the way they are built, there are some differences between them [10, 20].

Both nests have an outer shell made up of several protective layers, with secondary nests having between 5 and 10 layers, and primary nests having between 1 and 3 layers. The nest structures have a single entrance hole, which is at the bottom of the primary nests and at the side of the secondary nests, being much smaller in the last one (approx. 1.5 cm).

Primary nests are usually about 5–10 cm in diameter, and their shape has a usually circular form. Secondary nests are usually circular or oval in shape and can be as large as 60 x 80 cm [8, 10, 12].

Inside the nests, we can find the combs with the cells, where the queen lays her eggs. These combs are built by the workers as the colony grows, and the first combs are found in the upper part of the nest. In the embryonic nests, there are usually no more than three combs, and they can contain about 30 cells, while the secondary nests at the end of the cycle can have up to 10 combs with an average of 150 cells each and can house up to 1500 individuals [12, 17, 19].

As for their location, primary nests are constructed in places sheltered from the weather and possible predators and are mostly found in the eaves of windows, inside garages, shacks, under roofs or in very leafy bushes. Secondary nests are usually built outdoors at high altitudes in trees near a water source. However, they can also be observed in urban areas, in roof eaves, garages, empty hives and even on the ground [21–23].

The differences with the *Vespa crabro* nest must be focused on the secondary nest. The primary nests are very similar and can only be differentiated by identifying the insect inside. This species usually builds the secondary nest in holes in the bark of trees, holes dug in the ground or in cavities in buildings such as roofs. These are usually well camouflaged. Therefore, the shape of these nests depends on the location where they are built. There are occasions where they build them in plain sight, but they are usually in enclosed, sheltered places, such as warehouses, animal huts or garages. Another major difference with the *Vespa velutina* nest is the opening of the secondary nest, which is in the lower part of the nest and is large enough to allow the cells inside to be seen [18, 24].

5. Proliferation in Europe

The increase of global connections, especially in the commercial sphere, has resulted in a constant flow of people, goods and technologies. This movement has sometimes resulted in the accidental introduction of new species into countries where they did not previously exist. The adaptation of most invasive species to the new environment gives rise to a negative impact on ecosystems.

Regarding hornets, not all invasions are successful in establishing themselves in the invaded country, but there are exceptions. This is the case of the recent *Vespa velutina*, which was accidentally introduced into Europe in 2004, via the port of Bordeaux in cargo from China, which contained a dormant, fertilised queen [12, 25, 26]. Although the pathway of introduction remains somewhat uncertain, evidence that the spread into Europe began with a single queen was confirmed in 2022 by the research group of Dillane et al. [27]. It was not until a year later, in 2005, that the first record of the species was made in south-west France (Lot-et-Garonne). Since then, over the course of two decades, the insect has expanded its presence

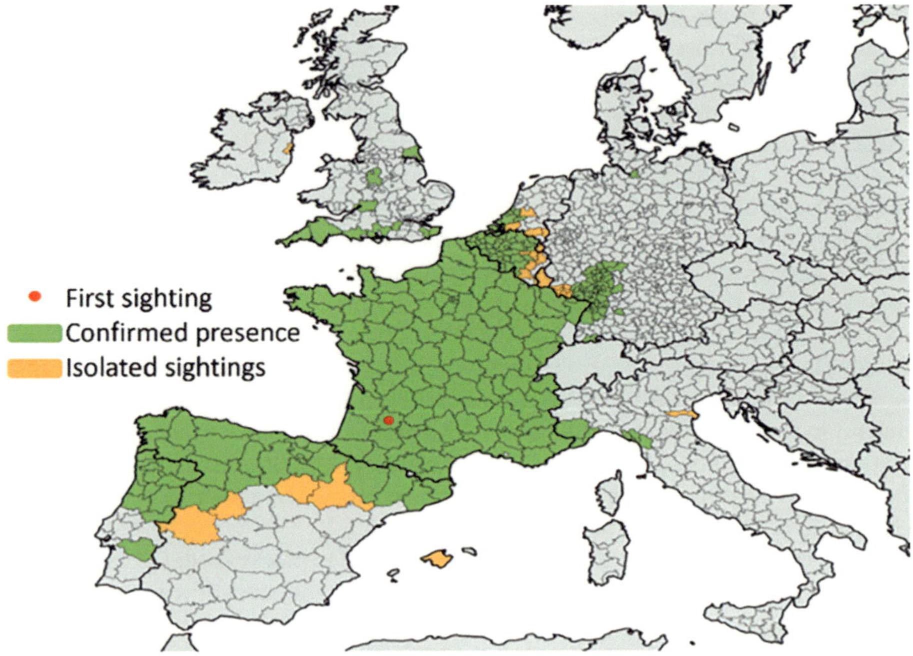

Figure 5.
Expansion of the Asian hornet in Europe in 2023. Created by https://www.mapchart.net/ (accessed on 9 January 2024). Figure from [11].

exponentially, expanding rapidly in Europe, with its presence confirmed in countries, such as Spain, Portugal, Belgium, the Netherlands, Germany, Italy, the United Kingdom and Ireland (**Figure 5**). Therefore, it represents the first successful invasion of the Vespidae family on this continent. It is estimated that *Vespa velutina* can extend its territory by 80 kilometres per year [13, 28]. The successful invasion and spread of the Asian hornet in Europe have been due to a combination of factors. The favourable climatic conditions have played a crucial role, being similar to their place of origin, with moderate temperatures and sufficient humidity. In addition, the inexhaustible source of available food and the absence of natural predators that could control their population have provided them with a suitable habitat to adapt and reproduce rapidly and successfully throughout Europe [29].

The massive attack on apiaries, fruit orchards and crops due to the need for protein and carbohydrates to feed larvae and adult hornets, respectively, is causing significant damage. These include crop losses, reduction in the number of pollinating insects and honey production, resulting in large economic losses in the agricultural and beekeeping sector, as well as in biodiversity [14, 15, 30]. In addition, *Vespa velutina* is also a danger to public health. Its sting can cause serious human health problems, especially for allergy sufferers. Although the attack on humans is scarce, they can occur if the colony feels threatened or if people come too close to their nests [12, 26, 31].

6. Control methods

The damage caused by *Vespa velutina* has led to the development of various control methods, with the aim of reducing its spread and minimising its impact.

The control methods developed up to now can be divided into those that inactivate the nest, those that trap hornets and those that help to minimise the impact on apiaries.

6.1 Inactivation of nests

This group of methods is based on eliminating or inactivating nests in order to minimise the damage that a nest can cause, as well as to prevent the hatching of new hornets, thus preventing the survival of the colony. It is worthwhile to consider that most of the methods in this category require specialised personnel, such as firefighters or forestry experts. In addition, the use of a specific *Vespa velutina* suit is mandatory for nest removal work. The costume is similar to the beekeeping costume but it is made with a stronger and thicker material, because of the larger size of sting of the hornets. Furthermore, it is necessary to wear special gloves and inside the hood, it is recommended to wear a helmet with a protective shield to avoid skin or eye contact with the biocide or venom. The latter is because the hornets spit venom when they feel threatened.

When the nest to be removed is primary, it can be easily removed without calling the professionals. This can be done by applying a common insecticide to the nest, by placing it in a container which must then be frozen to kill the individuals inside or by placing the nest in hot water (> 65°C) [10].

However, secondary nests must be removed by professionals due to the difficult access, where they usually build these nests, and/or the large size of the colony they contain. The most commonly used method to inactivate nests at high altitudes is by inoculation of a biocide with the help of a telescopic rod that can reach 3–25 m in length [10, 32]. In **Figure 6**, two workers can be seen using a telescopic stick to inactivate a secondary nest with biocide.

If the nest is in an area that cannot be reached with the telescopic pole and is not a danger to humans, it is usually not treated. However, if it does pose a danger, there are other methods of inactivation:

Figure 6.
Two workers inactivating a secondary nest of Vespa velutina *with a biocide applied by means of a telescopic stick. Source: Daniel Martínez.*

- Shotgun shots to destroy the nest (**Figure 7a**), the least reliable method, due to the high probability of colony survival, as the nest is broken up leaving surviving hornets that can rebuild the nest and continue to cause damage, even if the queen dies [10].

- Shooting with an adapted paintball gun using the system created by the company Asiatic Wasp Balls (Cantabria, Spain). It is based on projectiles to which the biocide chosen by the user is added and frozen until use. These are fired like any other projectile into the nest, but once they enter the nest they release the biocide as they thaw, killing both the individuals inside the nest and those that return later. However, sometimes, either due to lack of aim or because of the large size of the nest, it is necessary to fire several bullets, which can cause the nest to break up, and with it a high survival rate of the hornets. In addition, all bullets that do not hit the nest remain in the environment and can be ingested by other animals.

- Drones with a telescopic stick (**Figure 7b**) is very effective, but it is very expensive, and it is also necessary to have a clean flight path, because any brush with a tree branch can cause the drone to crash and break [33].

Within this group of methods, it is important to highlight the work carried out by Ruiz-Cristi et al., in which they have developed a new green control method for the inactivation of nests. Unlike traditional methods that inject different types of biocides, this method is based on the injection of hot steam inside the nest, reaching lethal temperatures for *Vespa velutina* [34].

Before carrying out any of these methods, it is recommended to make sure that the queen is inside the nest. The best time to do it is at night, when the entire colony is in the nest, and using only a red-light source as illumination, because *Vespa velutina* is not able to perceive it.

6.2 Trapping of hornets

The aim of individual trapping is to catch as many hornets as possible using attractive traps (**Figure 8**), usually made with sugar-rich ingredients. If the aim is to capture as many foundress queens and future queens as possible, to prevent growth

Figure 7.
On the left, a man shooting at a nest (source: Monica Torres) and on the right a drone with a stick attached inactivating a Vespa velutina nest (source: Miguel Souto).

Figure 8.
Attractive traps to control Vespa velutina. Source: Lioy S [35].

and the construction of new nests, these traps should be set in early spring and late autumn. This is the period of greatest effectiveness against the invasion, since one queen caught means one nest less.

It is important to highlight that this method is not selective for *Vespa velutina* and therefore attracts a multitude of insects, mainly Diptera and Vespidae.

Since the introduction of *Vespa velutina*, there has been an increase in the use of traps by private individuals, both commercial traps and traps made at home with natural ingredients, such as beer, white wine and sweet fruit juices. This uncontrolled trapping leads to a reduction of the local insect community and thus to possible consequences in future local biodiversity. Therefore, the placement of these traps in the Basque Country and in other regions of Spain must be requested and authorised by the competent local environmental institution [36].

For this selectivity problem, managing to create other types of traps or methods with more selective attractants for *Vespa velutina* is of special relevance for the control of this invasive species.

It is important to note that among the attractant traps are those made with *Vespa velutina* sex pheromones. Several research groups have worked on both the isolation of pheromones and the development of traps. The main objective of these traps is to capture sexually mature males, thus avoiding the mating of future queens [37–39].

6.3 Minimise the impact on apiaries

As is well known, the Asian hornet is a great predator of honeybees (*Apis mellifera*), for which *Vespa velutina* waits by gliding in front of the apiary, and when the bees are about to enter or leave it, they are easily hunted by their predator (**Figure 9**).

Because they only need the parts of the bees that are richest in protein to feed the larvae, adult hornets form a small protein pellet from the bee thorax, discarding all unnecessary body parts, such as the legs, wings, abdomen and head, and in some

Figure 9.
Vespa velutina attacking an apiary (left) and tearing apart a hunted bee before transporting it to the nest (right). Source: Jean Haxaire and Quentin Rome [40].

cases the cuticle of the thorax itself. If the apiary is weakened, the hornets are able to get inside and plunder it completely, thus obtaining the carbohydrates that the adults need to feed on the honey inside.

Due to the predatory behaviour towards bees and the absence of defensive techniques of bees against hornets, the application of specific control methods to treat apiaries is of great necessity.

Within this group of methods, the bee-hive muzzle and the electrified harp (**Figure 10**) stand out. Both are physical methods that are placed in the hive to block the entrance of hornets. The first is based on a net placed in the hive entrance, which has holes with a diameter that allows the bees to pass through without difficulty while blocking the entrance of hornets. Although it is a cheap method that can be constructed at home, it does not kill hornets [32].

The electrified harp is a structure that is placed between the hives, which can be square or rectangular, and which has wires in the middle connected to an electric current. These wires are placed at a distance from each other to allow only insects smaller than the *Vespa velutina* to pass through. When the hornets pass through this structure, they receive an electric shock. This is an effective and selective method of combating the species, however, on the downside, for treating apiaries with large numbers of hornets, it can be a great economic investment [32, 41].

A technique that combines the reduction of pressure in the apiaries with the inactivation of the nests is the placement of biocidal protein baits inside the nests [42, 43].

Figure 10.
Bee-hive muzzle (left) and electrified harp (right). Source: Lavignotte a. [32] and Rojas-Nossa, S. V. [41], Respectively.

The use of biocidal baits is a practice already used to combat other types of insect pests, such as Diptera, Hymenoptera, Hemiptera or Dictyoptera [44]. However, a few studies have been conducted on the use of protein baits to mitigate the destruction of bee apiaries by *Vespa velutina*.

The Basque Institute for Agricultural Research and Development, Neiker(Derio, Bizkaia) and the company supplying the baits, DTS OABE S.A. (Orozko, Bizkaia) carried out experiments to test the efficacy of these protein baits. These are placed near the apiaries to act as a protective barrier. The hornets are attracted to the pieces of bait that they take and introduce them into the nest to feed the larvae, and the effect of the biocide kills them by temporarily inactivating the nest, thus reducing the pressure on the apiary. In their research, they observed how the hornets effectively took the protein bait and flew away with it in their mouths (**Figure 11**). The presence of the bait caused the hornets to be attracted to them, allowing the bees to continue foraging.

Within 48 hours of the trial, a decrease in the number of hornets in the apiary was observed, maintaining a constant number of hornets for at least 2 weeks after treatment. However, after a few weeks, the number of hornets in the apiary began to increase. This is because the larvae that were operculated did not receive bait, hatching and continuing to grow the colony. Therefore, it is important to re-treat the apiary when an increase in the number of hornets is observed [36].

The protein baits used consist of different animal proteins and the insecticide phenylpyrazole, fipronil, as a biocide. This broad-spectrum insecticide is the used biocide in commercial protein baits [44]. It is important to consider that it would be necessary to study the possible risks that dead larvae and hornets and the remains of bait may present for birds, other insects and the environment.

Even knowing the proven effectiveness of bait in apiaries, due to the difficult access and search of nests in the wild because of the location of their constructions, it is not possible to know the effect caused by the bait on larvae and adults inside the nest. Studies with hornets under controlled environmental conditions in the laboratory represent a useful tool to obtain knowledge on the *Vespa velutina* in the nest.

Figure 11.
Asian hornets taking bait pieces from the study conducted by Neiker and DTS OABE S.A. in 2022.

7. *Vespa velutina's* behaviour in captivity

Very few studies in the literature have focused on the behaviour of *Vespa velutina* in captivity. The behavioural studies that have been reported are nest defence, colony activity and olfactory attraction [45–47].

In our research group, an observational study on the behaviour of *Vespa velutina* in captivity has been carried out [11]. Two secondary nests and one embryonic nest were captured and kept under controlled environmental conditions for up to 13 weeks for the secondary nest and 6 weeks for the embryonic nest. The different behaviours studied included adaptation to captivity, defence against disturbance, colony evolution and overwintering. In **Figure 12**, the locations of the different nests are given.

The nests were collected in transparent plastic boxes, which had little holes to allow hornets to breathe. An adhesive tape was used to close the boxes.

In order to be able to handle and introduce the *Vespa velutina* nests into the captive cage, the hornets were made to sleep using diethyl ether (99.7%) as an anaesthetic. Once all the hornets were asleep, they were placed, together with the nest, in a wooden cage (1 x 1 x 1 m) (**Figure 13**).

On each day, hornets were fed with a honey-water mixture (50:50), which was placed inside the cage and on the top grid impregnated in a common tissue and different pieces of fruit (banana, grape and apple) [16, 17, 20], which were placed on the side grids and inside the cage. Furthermore, fish protein, which is the main component of the tested bait pieces, was provided for the purpose of feeding the larvae. Both water and the honey-water mixture were supplied *ad libitum*, as reported in other studies [11, 17, 21].

To simulate outdoor conditions, the nests were adapted to a temperature-controlled environment (25–28°C) with light cycles (12 hours). In addition, different materials were provided to serve as supplementary food or to condition the nest (tree bark, fruit plants, stones and grass, pieces of nests) [20].

Figure 12.
Location of the Basque Country in Europe (left). Map of the Basque Country, with its three counties (right). The numbers indicate the locations of the Vespa velutina supplied nests. Image modified from paintmaps.com. Image from [11].

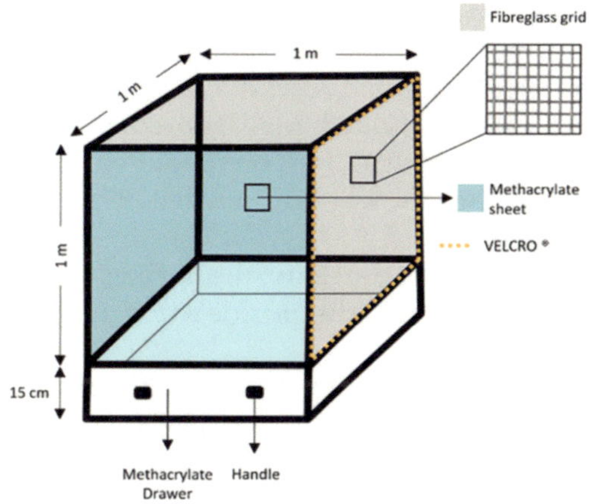

Figure 13.
Scheme of the cage used for the captivity study of Vespa velutina. Image from [11].

Each morning and evening, the different behaviours of the hornets were observed and recorded for at least one hour.

It must be considered that animals living and growing up in captivity experience drastically different conditions to those of the species in the wild, and therefore may act differently [23].

7.1 Behavioural observations in captivity

7.1.1 Adaptation to captivity conditions and defence against disturbance

After awakening from the anaesthetic, during the first two days, the hornets in the three nests showed aggressive behaviour towards their own colony and defensive against any kind of external disturbance. It was observed how they attacked and destroyed the nest, breaking the outer layers, fighting among themselves and even killing each other and extracting the larvae from the cells (**Figure 14**). Although this behaviour also occurred in the embryonic nest, the aggressiveness was less pronounced, especially with respect to the larvae. According to Perrad et al., this may be due to the fact that primary nests of *Vespa velutina* are not as aggressive towards

Figure 14.
Adult Vespa velutina hornet removing a larva from the nest, with another previously removed larva (left). Vespa velutina female hornets fighting with each other shortly after the nest was collected (right). Image slightly modified from [11].

humans, and therefore also towards their own colony, since colonies close to residential areas have been observed to become accustomed to the presence of humans [46].

Nervousness was another behaviour observed during the first days in captivity. In fact, when food and water were provided, they ignored it. This behaviour is not new to this study, as a study by Perrard et al. [46] on the activity of a colony kept in semi-captivity showed that when the nest was manipulated, it caused a disturbance during the first two days of captivity.

After approximately two days in captivity, the hornets showed a more relaxed and less aggressive behaviour towards the outside world and started to take food and drink water.

Regarding the protein food provided, the hornets did not take it on any of the occasions. This behaviour was related to the possible lack of live larvae inside the nest or due to the fact of being out of their habitat, which could have affected them when taking care of the larvae.

However, due to the absence of scratching noises inside the nest, which are characteristic of larvae when begging for food, the most reliable hypothesis for not taking the protein food is the lack of live larvae inside the nest [46].

Due to the lack of live larvae, it was not possible to obtain information on what happens in the nest when bait rich in protein is introduced.

However, in the embryo nest, the survival of the larvae and how they were fed with protein food by the workers could be verified. These results provide an opportunity to carry out *in vivo* studies using embryo nests to evaluate the effectiveness of protein baits in the control of this invasive species.

7.1.2 Evolution of the colony and overwintering

Over the first two weeks, the operculated cells in the nest reception area began to hatch, and new hornets, mostly males, emerged from these cells. The new individuals were identified by their characteristic whitish colouring on the thorax and wings attached to the body, which occurs during the first days after hatching, due to a lack of keratin. As expected, due to the period of the life cycle in which the embryo nest was located, most of the newborns were females.

As the weeks went by, the number of new hornets decreased. In the research carried out by Monceau et al. [47], this evolution was also observed, detailing when the queen is caging it stopped laying more eggs. It is thought that this could be due to a lack of nutrients or resources to regulate the colony or its sanitary maintenance, as well as the stress generated by captivity.

The separation of individuals in the colony by sex was another of the behaviours observed, as occurs in wildlife. With the sexual maturity of the males, they were placed in the corner of the cage furthest away from the nest. In nest 1, when males approached the nest, several female hornets were observed to act aggressively towards them, attacking them with bites and stings to the point of death. Following these attacks, dismembered male hornets were observed in the cage. This type of reaction, previously observed and reported in *Vespa simillima* (Cameron 1903), could be due to a mechanism of females to avoid inbreeding [47].

This behaviour was not observed in the embryonic nest, as only three males hatched and died the following day.

Mating between the hornets in nest 1 was detected up to 5 times between the fourth and sixth week of captivity (**Figure 15**). As in the wild, mating took place on the cage walls furthest away from the nest.

Figure 15.
A couple of Vespa velutina hornets from nest 1 mating in captivity. Image from [11].

With the passing of the weeks, in both secondary nests the number of individuals in the colony started to decrease, with new dead hornets appearing every day. This started to occur from weeks 7 and 10 for nests 1 and 2, respectively. When the observed number of live hornets in the cage was less than 15 and females were present, in order to simulate the arrival of autumn and the entry into dormancy of the future queens, the laboratory temperature was lowered (from 25–28°C to 16–17°C) and food was gradually reduced, until there was no food.

Following some weeks under non-feeding conditions and low temperatures, in nest 1, it was observed that only three female hornets had survived, so they were considered as possible future fertilised queens. Therefore, in order to try to complete the life cycle of *Vespa velutina* in captivity, a hibernation study was carried out. For this purpose, the environmental conditions were as follows. The temperature was lowered to 13°C and the light cycle was removed, simulating the hiding place they seek for the winter.

To ensure that hibernation was not interfered with, the condition of the hornets was checked once a week without turning on the light. **Figure 16** shows how during

Figure 16.
Vespa velutina hornets from nest 1 in state of lethargy in captivity. Image from ref. [11].

one of the weeks of the hibernation study, one of the hornets shifted from the corner and moved slightly from its initial position, while the other one and the one on the branch did not move.

After 5 weeks of hibernation, the hornets awoke from dormancy. This made it necessary to start a new biological cycle, for which the environmental conditions had to be modified, increasing the temperature (21°C), repeating the light cycles, and supplying food and water. The hornets were active again, eating and flying normally. However, after one week, all three hornets were found dead. No conclusion could be drawn as to why this happened. Different factors such as the environment, which is not ideal to start the cycle again, or the existence of other possible fertilised queens in the same environment, as in nature, for survival reasons, they do not hibernate in groups and build the primary nest individually far away from other gynes.

8. Conclusions

Vespa velutina nigrithorax is considered as invasive species in Europe and represents a threat to the biodiversity, agriculture, apiculture and human health. Its expansion in Europe is estimated at up to 80 km/year.

The use of protein baits with fipronil in apiaries when in the presence of *Vespa velutina* can reduce the pressure of the hornets on the hives for at least two weeks after baiting. Furthermore, bees can recover their normal activity.

Apart from the known control methods, more selective and environmentally friendly methodologies are being developed to fight against this species.

The observational study of *Vespa velutina*'s behaviour in captivity has showed the possibility to maintain secondary nests up to 13 weeks under controlled environmental conditions.

The captivity studies allow deepening the knowledge surrounding *Vespa velutina's* ethology, obtaining useful information on the behaviour of this invasive species.

Acknowledgements

Omaira de la Hera thanks the University of the Basque Country (UPV/EHU) for her predoctoral grant. Authors thank the Basque Country Government (projects PUE 2018_1_0007 and PUE 2021_1_008) and the University of the Basque Country (UPV/EHU) (project US21/35) for financial support. We thank the public company BASALAN and Avispa Asiatica Association (Villarcayo, Spain) for providing nests for the laboratory tests.

Conflict of interest

The authors declare no conflict of interest. The funders had no role in the design of the study; in the collection, analyses or interpretation of data; in the writing of the manuscript; or in the decision to publish the results.

DOI: http://dx.doi.org/10.5772/intechopen.1004942

Author details

Omaira de la Hera, Maria Luz Alonso and Rosa Maria Alonso*
Faculty of Science and Technology, Analytical Chemistry Department, FarmArteM Group, University of the Basque Country (UPV/EHU), Bilbao, Biscay, Spain

*Address all correspondence to: rosamaria.alonso@ehu.eus

References

[1] Archer ME. A phylogenetic study of the species of the genus Vespa (Hymenoptera: Vespinae). Entomology Scandinavian. 1994;**24**:469-478

[2] Nguyen LTP, Saito F, Kojima JI, Carpenter JM. Vespidae of Viet Nam (Insecta: Hymenoptera) 2. Taxonomic notes on Vespinae. Zoological Science. 2006;**23**:95-104. DOI: 10.2108/zsj.23.95

[3] Perrard A, Pickett KM, Villemant C, Kojima JI, Carpenter J. Phylogeny of hornets: A total evidence approach (Hymenoptera, Vespidae, Vespinae, Vespa). Journal of Hymenoptera Research. 2013;**32**:1-15. DOI: 10.3897/JHR.32.4685

[4] Matsuura M, Yamane S. Biology of the Vespine Wasps. Berlin, Germany: Springer-Verlag; 1990

[5] Marris G. The Asian hornet: Part 1. Bee Craft. September 2011:16-18

[6] Pérez-De-Heredia I, Darrouzet E, Goldarazena A, Romón P, IturrondobeitiaJC.Differentiatingbetween gynes and workers in the invasive hornet Vespa velutina (Hymenoptera, Vespidae) in Europe. Journal of Hymenoptera Research. 2017;**60**:119-133. DOI: 10.3897/jhr.60.13505

[7] Rome Q, Muller FJ, Touret-Alby A, Darrouzet E, Perrard A, Villemant C. Caste differentiation and seasonal changes in Vespa velutina (Hym.: Vespidae) colonies in its introduced range. Journal of Applied Entomology. 2015;**139**:771-782. DOI: 10.1111/jen.12210

[8] Bunker S. The Asian Hornet Handbook. Mountain View, CA, USA: Psocid Press; 2019

[9] Stephen M. The Asian Hornet: Threats, Biology & Expansion. Mytholmroyd, UK: IBRA and Northern Bee Books; 2017

[10] Galartza E. Manual para la gestión de la Avispa asiática (Vespa velutina) 2016. Medio Natural. 2016:1-81

[11] de la Hera O, Alonso ML, Alonso RM. Behaviour of Vespa velutina nigrithorax (Hymenoptera: Vespidae) under controlled environmental conditions. Insects. 2023;**59**:1-14. DOI: 10.3390/insects14010059

[12] Chauzat M-P, Martin S. A foreigner in France: The Asian hornet. Biologist. 2009;**2**:86-91

[13] Monceau K, Bonnard O, Thiéry D. Vespa velutina: A new invasive predator of honeybees in Europe. Journal of Pest Science. 2014;**87**:1-16. DOI: 10.1007/s10340-013-0537-3

[14] Leza M, Herrera C, Marques A, Roca P, Sastre-Serra J, Pons DG. The impact of the invasive species Vespa velutina on honeybees: A new approach based on oxidative stress. Science of the Total Environment. 2019;**689**:709-715. DOI: 10.1016/j.scitotenv.2019.06.511

[15] Lioy S, Bergamino C, Porporato M. The invasive hornet Vespa velutina: Distribution, impacts and management options. CABI Reviews. 2022:1-14. DOI: 10.1079/cabireviews202217030

[16] Abe T, Tanaka Y, Miyazaki H, Kawasaki YY. Comparative study of the composition of hornet larval saliva, its effect on behaviour and role of trophallaxis. 1991;**99**:79-84. DOI: 0306492/913.00 0.00 0 1991

[17] Villemant C, Barbet-Massin M, Perrard A, Muller F, Gargominy O, Jiguet F, et al. Predicting the invasion risk by the alien bee-hawking yellow-legged hornet Vespa velutina nigrithorax across Europe and other continents with niche models. Biological Conservation. 2011;**144**:2142-2150. DOI: 10.1016/j.biocon.2011.04.009

[18] Erturk O, Bagdatli E. A comprehensive study on nest materials of Vespa crabro and Polistes dominula: Chemical properties and biological characterization with antioxidant and antimicrobial activity. Biologia (Bratisl). 2019;**74**:797-812. DOI: 10.2478/s11756-019-00209-y

[19] Sánchez XF, Charles RJ. Notes on the nest architecture and colony composition in winter of the yellow-legged Asian hornet, Vespa velutina Lepeletier 1836 (Hym.: Vespidae), in its introduced habitat in Galicia (NW Spain). Insects. 2019;**10**:237, 1-17. DOI: 10.3390/insects10080237

[20] Crespo N, Louzada J, Fernandes LS, Tavares PB, Aranha J. Microscopic identification of anatomical elements and chemical analysis of secondary nests of Vespa velutina nigrithorax du Buyson. Insects. 2022;**13**:537, 1-16. DOI: 10.3390/insects13060537

[21] Herrera C, Marqués A, Colomar V, Leza MM. Analysis of the secondary nest of the yellow-legged hornet found in the Balearic Island reveals its high adaptability to Mediterranean isolated ecosystems. Island Invasives: Scaling up to Meet the Challenge. 2019;**62**:375-380

[22] Kennedy PJ, Ford SM, Poidatz J, Thiéry D, Osborne JL. Searching for nests of the invasive Asian hornet (Vespa velutina) using radio-telemetry. Communications Biology. 2018;**1**:88, 1-8. DOI: 10.1038/s42003-018-0092-9

[23] Monceau K, Thiéry D. Vespa velutina nest distribution at a local scale: An 8-year survey of the invasive honeybee predator. Insect Sci. 2017;**24**:663-674. DOI: 10.1111/1744-7917.12331

[24] Cini A, Cappa F, Petrocelli I, Pepiciello I, Bortolotti L, Cervo R. Competition between the native and the introduced hornets Vespa crabro and Vespa velutina: A comparison of potentially relevant life-history traits. Ecological Entomology. 2018;**43**:351-362. DOI: 10.1111/een.12507

[25] Villemant C, Haxaire J, Streito J. Premier bilan de l'invasion de Vespa velutina Lepeletier en France (Hymenoptera, Vespidae). Bulletin de la Société entomologique de France. 2006;**111**:535-538

[26] Schwartz C, Villemant C, Rome Q, Muller F. Vespa velutina (frelon asiatique) : Un nouvel hyménoptère en France. Revue Française d'Allergologie. 2012;**52**:397-401. DOI: 10.1016/j.reval.2012.06.002

[27] Dillane E, Hayden R, O'Hanlon A, Butler F, Harrison S. The first recorded occurrence of the Asian hornet (Vespa velutina) in Ireland, genetic evidence for a continued single invasion across Europe. Journal of Hymenoptera Research. 2022;**93**:131-138. DOI: 10.3897/jhr.93.91209

[28] Bordier C. The spread of Vespa velutina, the invasive predator of honeybees in Europe. Insectes Sociaux. In: 30é congress UIEIS Section francaise. Avignon, France; 2019

[29] Otis GW, Taylor BA, Mattila HR. Invasion potential of hornets (Hymenoptera: Vespidae: Vespa spp.). Frontiers in Insect Science. 2023;**3**:1-21. DOI: 10.3389/finsc.2023.1145158

[30] Fortuna TM, Le Gall P, Mezdour S, Calatayud PA. Impact of invasive insects on native insect communities. Current Opinion in Insect Science. 2022;**51**:100904. DOI: 10.1016/j.cois.2022.100904

[31] Feás X. Human fatalities caused by hornet, wasp and bee stings in Spain: Epidemiology at state and sub-state level from 1999 to 2018. Biology (Basel). 2021;**10**:1-13. DOI: 10.3390/biology10020073

[32] Turchi L, Derijard B. Options for the biological and physical control of Vespa velutina nigrithorax (Hym.: Vespidae) in Europe: A review. Journal of Applied Entomology. 2018;**142**:553-562. DOI: 10.1111/jen.12515

[33] Lee CG, Yu SH. Exterminator for the nests of Vespa velutina nigrithorax using an unmanned aerial vehicle. Drones. 2023;7:281, 1-12. DOI: 10.3390/drones7040281

[34] Ruiz-Cristi I, Berville L, Darrouzet E. Characterizing thermal tolerance in the invasive yellow-legged hornet (Vespa velutina nigrithorax): The first step toward a green control method. PLoS One. 2020;**15**:10, 1-14. DOI: 10.1371/journal.pone.0239742

[35] Lioy S, Laurino D, Capello M, Romano A, Manino A, Porporato M. Effectiveness and selectiveness of traps and baits for catching the invasive hornet vespa velutina. Insects. 2020;**11**:1-13. DOI: 10.3390/insects11100706

[36] Barandika JF, de la Hera O, Fañanás R, Rivas A, Arroyo E, Alonso RM, et al. Efficacy of protein baits with Fipronil to control Vespa velutina nigrithorax (Lepeletier, 1836) in apiaries. Animals; 2023;**13**:2075, 1-15. DOI: 10.3390/ani13132075

[37] Nieh J. Luring Hornets: Scientists Unlock Sex Pheromone of Notorious Honey Bee Predator. 2017. Available from: https://PhysOrg/News/2017-10-Luring-Hornets-Scientists-Sex-PheromoneHtml

[38] Wen P, Cheng YN, Dong SH, Wang ZW, Tan K, Nieh JC. The sex pheromone of a globally invasive honey bee predator, the Asian eusocial hornet, Vespa velutina. Scientific Reports. 2017;**7**:1-11. DOI: 10.1038/s41598-017-13509-7

[39] Couto A, Monceau K, Bonnard O, Thiéry D, Sandoz JC. Olfactory attraction of the hornet Vespa velutina to honeybee colony odors and pheromones. PLoS One. 2014;**9**:12, 1-19. DOI: 10.1371/journal.pone.011594

[40] Rome Q, Perrard A, Muller F, Villemant C. Monitoring and control modalities of a honeybee predator, the yellow-legged hornet Vespa velutina nigrithorax (Hymenoptera:Vespidae). Aliens. 2011;**31**:7-13

[41] Rojas-Nossa SV, Dasilva-Martins D, Mato S, Bartolomé C, Maside X, Garrido J. Effectiveness of electric harps in reducing Vespa velutina predation pressure and consequences for honey bee colony development. Pest Management Science. 2022;**78**:12, 5142-5149. DOI: 10.1002/ps.7132

[42] Spurr EB. Protein bait preferences of wasps (Vespula vulgaris and v. germanica) at Mt Thomas, Canterbury, New Zealand. N Z. Journal of Zoology. 1995;**22**:281-289. DOI: 10.1080/03014223.1995.9518043

[43] Peters BC, Wibowo D, Yang GZ, Hui Y, Middelberg APJ, Zhao CX. Evaluation of baiting fipronil-loaded silica nanocapsules against termite colonies in fields. Heliyon. 2019;**5**:1-10. DOI: 10.1016/j.heliyon.2019.e02277

[44] Edwards E, Toft R, Joice N, Westbrooke I. The efficacy of Vespex®

wasp bait to control Vespula species (Hymenoptera: Vespidae) in New Zealand. International Journal of Pest Management. 2017;**63**:266-272. DOI: 10.1080/09670874.2017.1308581

[45] Van Itterbeeck J, Feng Y, Zhao M, Wang C, Tan K, Saga T, et al. Rearing techniques for hornets with emphasis on Vespa velutina (Hymenoptera: Vespidae): A review. Journal of Asia-Pacific Entomology. 2021;**24**:103-117. DOI: 10.1016/j.aspen.2021.03.009

[46] Perrard A, Haxaire J, Rortais A, Villemant C. Observations on the colony activity of the asian hornet vespa velutina lepeletier 1836 (hymenoptera: Vespidae: Vespinae) in France. Annales de la Societe Entomologique de France. 2009;**45**:119-127. DOI: 10.1080/00379271.2009.10697595

[47] Monceau K, Bonnard O, Thiéry D. Relationship between the age of Vespa velutina workers and their defensive behaviour established from colonies maintained in the laboratory. Insectes Sociaux. 2013;**60**:437-444. DOI: 10.1007/s00040-013-0308-4

Chapter 9

Physiological and Environmental Impact of Temperature Change on Bumblebee Flight

Isa Gan, Munjal Shah, Javid Bayandor and Francine Battaglia

Abstract

Insects are impressive flyers due to their intricate wing anatomy, exceptional maneuvering abilities, and capacity to fly in harsh conditions. Bees adapt to extreme environmental conditions through thermoregulation, which allows them to lose or gain heat to regulate their body temperature and increase survivability and foraging capability. The temperature of the thorax, which is regulated by heat transfer between the body and the environment must be kept within a specific range to generate enough power to fly. Computational studies examining heat transfer effects on the aerodynamic performance of bees are limited. This study analyzes effects of ambient temperature and heat dissipation from the body on lift and thrust using morphologically accurate models from *Bombus pensylvanicus* scans. Three-dimensional incompressible Navier-Stokes equations were solved to predict flow around the bee in different environments. The results of the study showed that even though the thoracic activity changes the flow conditions around the wing, the effect of heat transfer given the assumptions made was not found to have a significant effect. Ambient conditions, however, play a crucial role in flight performance. Increasing ambient temperature reduced the pressure surrounding the wings, which led to decreased lift and aerodynamic power production at hovering and forward flight.

Keywords: aerodynamics, *Bombus pensylvanicus*, computational fluid dynamics, environmental conditions, flapping flight, thermoregulation

1. Introduction

Miniaturizing flying vehicles has become a trend with recent technological progress; hence, insect flight has fascinated researchers due to the complex lift generation mechanism relative to the insect size. Dickinson et al. [1] conducted experiments to investigate the unsteady mechanism of insects using a *Drosophila* wing. The results showed that the translational mechanism or delayed stall had a vital contribution to lift generation; however, it was inadequate to keep insects aloft. It was noted that rotational circulation and wake capture mechanisms also played an essential role in the aerodynamics of insect flight. Nagai et al. [2] compared experimental and numerical results using a dynamically-scaled bumblebee in hovering and forward flight and

 IntechOpen

concluded that the experimental and numerical results agreed and delayed stall, rotational circulation and wake capture mechanisms were functional; nonetheless, they took place in different ways during the downstroke and upstroke of the wing. Lyu et al. [3] computationally investigated the effect of the Reynolds number on the flight of an assumed rigid flat wing planform of a dronefly (*Eristalis tenax*) under a very low Reynolds number ($Re < 100$). The study showed that tiny insects must have different flapping patterns than bigger insects due to stronger viscous forces that decrease lift and thrust for $Re < 70$.

To better understand the importance of wing morphology on bee aerodynamic performance, Feaster et al. [4] performed two-dimensional computational simulations to compare a morphologically-accurate model of the bumblebee wing with flat plate and ellipse wings. Their results showed that the wing corrugation is a critical factor because the corrugated wing generates shorter and several leading-edge vortex (LEV) structures than the flat plate and the ellipse during downstroke, thereby leading to elevated lift production of the corrugated wing. Shah et al. [5, 6] considered the effect of wing corrugation in three-dimensional numerical simulations using corrugated and smooth wings of *Bombus pensylvanicus* in hovering and forward velocities. The results proved that the morphologically-accurate model increased lift by 7–12% and decreased thrust by 49% at the higher forward speeds (2.5 and 4.5 m/s) as there were no major differences during hovering. Flexible wings are able to control and direct flow, thus leading to a high lift-to-drag ratio compared to fixed wings [7]. Tobing et al. [8] computationally analyzed the effect of wing flexibility on the aerodynamics of bumblebees. The results illustrated that the flexible wing provides more stability than the rigid wing due to preventing LEV separation, resulting in obtaining 30% higher lift than the rigid wing.

Bumblebees can fly in cold and warm weather conditions due to their adaptation skills to changing air conditions. Even though they succeed in flying in harsh conditions, their flight performance is limited by the environment [9]. In the study by Heinrich [10, 11], it was reported that the body temperature of bumblebees correlated with ambient temperature and the excessive heat in the thorax due to the high air temperature limited the bee foraging time; thus the bees must transfer heat. Heinrich and Kammer [12] conducted an experimental study on bumblebees about the warming-up of flight muscles and thoracic temperature stabilization of bumblebees. Their results showed that bumblebees must warm up their flight muscles before flying by employing wing-beat until the thoracic temperature is over 30°C and the heat production in the thorax rises with increasing wing-beat frequency. Moreover, the excessive heat in the thorax was transferred to the abdomen and was dissipated by convective heat transfer to stabilize the amount of heat in the thorax. Roberts and Harrison [13] predicted that increasing ambient temperature decreases metabolic rate and the wing-beat frequency, thereby causing lower aerodynamic power production.

Bumblebees inhabit many environments and forage long distances at different temperatures due to their adaptation skill to changing environments by employing heat transfer between body and ambient. How does heat transfer affect their flight and how do they respond to ambient temperature changes? To address these questions, the current study was performed to reveal the thermal impacts on the aerodynamics of insects using computational fluid dynamics (CFD). A morphologically-accurate model of a bumblebee (*Bombus pensylvanicus*) was used as the computational model. In contrast to experimental studies, computational studies to investigate the effect of thermal properties on bees are limited. The motivation of the current study was to present a computational approach to investigate the thermal response of bumblebee flight in hovering and forward flight.

The study will investigate both the effect of the thoracic activity of bees that results in heat transfer with the surroundings and the impact of ambient temperature changes on the flight performance of bumblebees. Due to the correlation of the body temperature of bees with ambient temperature, different body and ambient temperatures are specified using the experimental data of Heinrich [10, 11]. The effects of heat transfer between the body and ambient temperature on the aerodynamics of bumblebees will be analyzed computationally.

2. Bumblebee

Bees are classified into various groups of insects. There are a variety of bee species that live in a colony with a thousand individuals. Bees inhabit diverse environments ranging from cold arctic regions to humid tropical rain forests to hot deserts due to their superior environmental adaptation [14]. They play a critical role in the pollination of wild plants, flowers and harvests, supporting a balance in Nature as well as economic acquisition [15]. Their survivability, however, highly depends on environmental conditions in their habitat [16]. To tolerate unwanted climate conditions, unique adaptation skills are utilized to maintain their life cycle.

Bumblebees with 300 species are among 20,000 other bee species and their mass ranges from 65 mg for the smallest worker to 830 mg for the largest queen species [17, 18]. They inhabit cold and hot environments and are key species for pollinating wild herbs, crops, and vegetables. Unlike honeybees and other bee species, they forage at lower ambient temperatures and for long hours, carry pollen that is nearly 80% percent of their body weight, and are agile fliers that make complex maneuvers [5, 11].

2.1 Morphology

Morphological characteristics play an important role in understanding the flight mechanics of flying creatures. These parameters are grouped into gross and shape parameters. Gross parameters that are a raw description of insect morphology include a body for an assigned length and added mass for a given wing length, as the shape parameters are the distribution of the gross parameters such as radii of wing area, moment of wing area, and body and wing shapes, etc. [19].

The wing shape and body mass determine the flight style of a flying creature. High aspect-ratio wings allow them to perform complex maneuvers and waste less energy; however, this is not purely explained in terms of aspect-ratio [20]. Other wing parameters like length, wing area, flexibility, body and virtual mass, and wing loading have notable impacts on the efficiency of flight performance of flying animals. The wing flexibility helps the aerodynamic performance and is superior compared with rigid wings. Insects utilize flexible wings, and the flexibility provides delayed stall, increased lift, and decreased drag [7, 21, 22].

The wing morphology can determine the optimal wing motion of bees [23]. The bumblebee wing is comprised of veins and cells (**Figure 1(a)**), and the shape of the wing differs among bumblebee species. Also, their wings are small and require rapid beats to keep them aloft [5, 18, 24]. Corrugated wings (**Figure 1(b)**) contribute to the aerodynamic performance and flight stability of the bees [25, 26]. It helps larger LEV formation during the downstroke, thus leading to lift augmentation by extending suction pressure on the wing dorsal side [5].

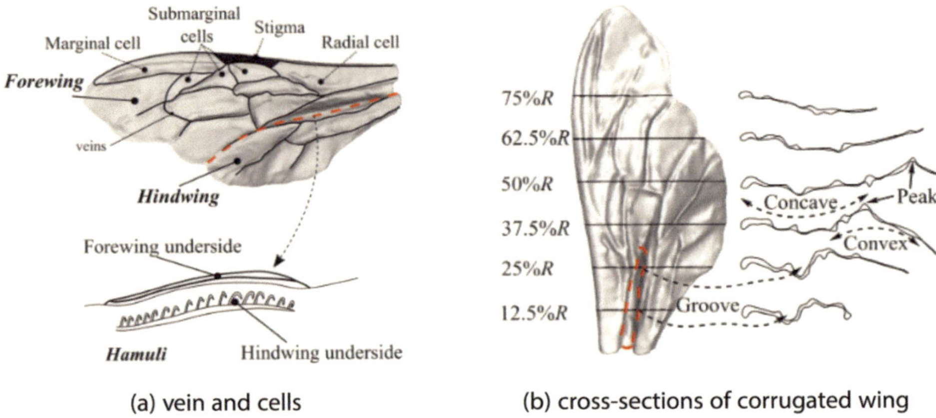

Figure 1.
*The wing of a bumblebee (*Bombus pensylvanucus*), adopted from Ref. [5].*

2.2 Thermoregulation

Bees need sufficient fuel sources and a stable thoracic temperature range to maintain their ability to forage in all weather conditions [10]. There are diverse mechanisms such as hair density, hair color, and heat transfer with environment to increase their survivability due to the fact that different species adapt to their unique habitats [14]. The most significant adaptation is the mechanism of heat loss or gain, named thermoregulation, which tolerates harsh environmental conditions [13]. The contraction of the flight muscles to initiate flight generates power and heat in the thorax (see in **Figure 2**). The flight muscle temperature must be regulated to maintain required work output from bees [10]. Thermoregulation is an adaptation to severe environmental conditions, increasing the bee survivability and ability to forage long distances efficiently [16].

The hair coats of bumblebees shown in the left half of **Figure 2** provide excellent insulation to limit convective heat loss [27]. Nixon and Hines [16] performed a study to examine the effect of the physical features of bumblebees on their temperature regulation. It is reported that the color and pile length played key roles in the passive heating of bumblebees, increasing their survivability in different environments.

The aerodynamic performance of insects is dependent on the body and ambient temperatures [11]. During the flight, muscles are heated by the wing motion and positively correlate with increasing ambient temperature. However, low thorax

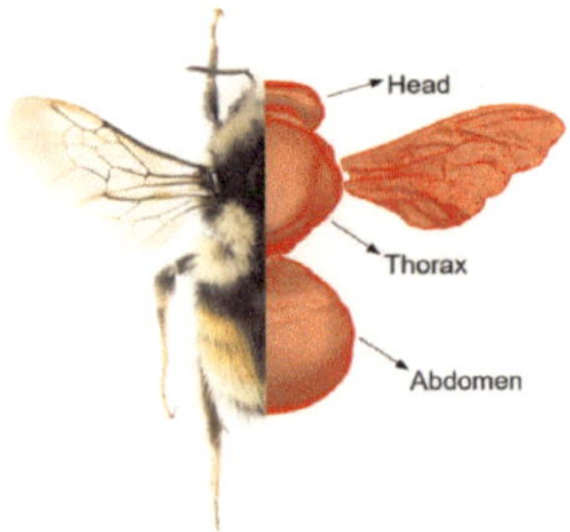

Figure 2.
Bombus vancouverensis *with the morphological accurate computational model of the American bumblebee (*Bombus pensylvanucus*) on the right half. The bumblebee image on the left half is credited to ref. [16].*

temperature restrains power production, as does high temperature, because bees cannot heat the flight muscles at low temperatures. Maintaining thoracic temperatures within a limited range is a vital process to maintain the required power output during flight [10, 28, 29].

Bumblebees require thoracic temperatures within $30°C - 44°C$ to generate the needed power output to rise from the ground and forage [18]. Regardless of body mass, thoracic temperature controls bee muscle functions. As observed by Heinrich [18], the majority of bumblebees cannot maintain flight below $10°C$ air temperature, whereas large bees or queens are able to fly at lower ambient temperatures down to 0° C. To keep the thorax temperature during flight at low and high ambient temperatures, different methods are used to heat up and keep the body warm.

Shivering is a method to raise the temperature level of flight muscles that comprise the larger part of the thorax with a slight contraction of flight muscles before the flight takes place [10, 17]. Most bee species need to warm-up their flight muscles for some tasks including before taking off, pollen collection, cooling and warming up of clusters [30].

During long flight periods, bees transfer heat between their thorax and abdomen, and even ambient to keep the thoracic temperature within optimal conditions. The abdomen plays an important role in the heat balance of bees. Most bees are capable of losing heat by transferring to the abdomen from the thorax [31]. The experiments by Church [27, 32] showed that heat dissipation by convection is a more effective way than heat loss by evaporation of the body water in most small insects. Bumblebees only perform cooling with evaporation when they are compelled to fly at high ambient temperatures [31, 33]. Ambient temperatures can negatively affect bee thoracic activity and their aerodynamic performance. The metabolic rate of bees reduces with increasing air temperature, resulting in elevated thoracic temperature [34, 35]. Moreover, a shift in ambient temperature also leads to decreasing aerodynamic performance due to the density reduction and viscosity increase of the air, requiring bees to increase wing-beat frequency to compensate for the aerodynamic loss [10, 13, 16].

3. Numerical methodology

The flow behavior around the bee is simulated using computational models of the unsteady Navier-Stokes equations. The Navier-Stokes equations are a mathematical representation of Newton's second law of motion relating forces to fluid momentum. The forces account for surface stresses due to fluid gradients in pressure and fluid viscosity. The set of partial differential equations are known as conservation of momentum and are coupled to an equation for conservation of mass. Together, the solution provides information for how pressure and fluid viscosity affect the fluid motion. When temperature changes are important, an equation for conservation of energy is solved to provide information how the fluid motion is affected by temperature gradients. Thus, the time-dependent motion of flapping bumblebee wings is impacted by the thoracic temperature, air temperature and surrounding air pressure as the bee flies, and can be understood by solving the following equations. The air flow is assumed to be laminar, viscous and incompressible.

The equation for conservation of mass is:

$$\frac{\partial \rho}{\partial t} + \vec{\nabla} \cdot \left(\rho \vec{v} \right) = 0 \qquad (1)$$

where ρ is density, t is time, and $\vec{v}$ is the velocity vector.
The momentum conservation equation is:

$$\frac{\partial}{\partial t}\left(\rho\vec{v}\right) + \vec{\nabla}\cdot\left(\rho\vec{v}\vec{v}\right) = -\vec{\nabla}p + \vec{\nabla}\cdot\overline{\tau} + \rho\vec{g} \qquad (2)$$

where p is pressure, $\overline{\tau}$ is the viscous stress tensor and $\vec{g}$ is gravitational vector.
The fluid viscous stress tensor is defined as:

$$\overline{\tau} = \mu\left[\left(\vec{\nabla}\vec{v} + \vec{\nabla}\vec{v}^{T}\right) - \frac{2}{3}\vec{\nabla}\cdot\vec{v}\right] \qquad (3)$$

where μ is the dynamic (absolute) viscosity of the fluid.
The energy conservation equation is:

$$\frac{\partial}{\partial t}(\rho E) + \vec{\nabla}\cdot\left((\rho E + p)\vec{v}\right) = \vec{\nabla}\cdot k\nabla T + \vec{\nabla}\cdot\left(\overline{\tau}\cdot\vec{v}\right) \qquad (4)$$

where E is total energy, and k is the conduction coefficient term.

The computational study was performed using a pressure-based solver in a fully coupled scheme. The implicit discretization was applied to the governing equations. For the gradient of the variables, the least squares cell-based method was selected. To obtain accurate results, the second-order upwind scheme was chosen. To maintain the stability of the simulation, the time step was determined by using the Courant-Friedrichs-Levy (CFL) number:

$$CFL = \frac{U_0\Delta t}{\Delta x} \qquad (5)$$

where U_0 is the reference velocity on the wing, Δt is the time step and Δx is the average mesh cell size. Transient simulations were carried out for 5 cycles.

3.1 Validation of flapping wing kinematics

The validation of the numerical setup was performed using the data of a fruit fly experiment conducted by Dickinson et al. [1]. The kinematic pattern of the flapping wing was a sinusoidal motion for stroke position and angle of attack. A dynamically-scaled model of a fruit fly (*Drosophila melanogaster*) was built to imitate the kinematics of *Drosophila* wings and obtain aerodynamic forces over the wing using sensors. In the experiment, the motion of two wings was powered by computer-controlled motors and gearboxes attached to the wings. Furthermore, to reduce the wall effects, a 1 m $\times$ 1 m $\times$ 2 m tank was designed in which the wings were plunged in mineral oil with a density of 880 kg/m ^{3} and kinematic viscosity of 115 cSt.

The *Drosophila* wing used in the experiment was constructed from 25 cm long and 3.2 mm thick Plexiglas. The wing rotation axis was $0.2c$ from the leading edge, where c is the chord length. The wing kinematics was described by using stroke amplitude, $\phi = 160°$, angle of attack, $\alpha = 40° \pm 5°$ at mid-stroke for both downstroke and upstroke, and wing-beat frequency, $f = 145$ mHz.

In the numerical study, the domain size, wing morphology, fluid properties and wing kinematics were consistent with those in the experimental study. The wing, overset and computational setup are shown in **Figure 3**, where symmetry is assumed. Background ((**Figure 3(a)**) and overset (**Figure 3(b)**) domains include 770 k and 1.4 m polyhedral cells, respectively. During the wing movement, the stroke angle changed:

$$\phi(t) = \frac{\phi_{max}}{1.7424}\left[1 - \left(2\cos^{-1}(0.98\sin(2\pi ft))\right)\right] \quad (6)$$

where $\phi_{max} = 160°$ was the maximum stroke angle. The angle of attack of the wing varies:

$$\alpha(t) = \frac{\alpha_{max}}{\tanh C_\theta}[-\tanh(C_\theta \cos(2\pi ft))] \quad (7)$$

where $\alpha_{max} = 40°$ at mid-stroke and $C_\theta = 2.3$. The parameter C_θ determined the function deviation from sine to square waveform. The function transformed into a square wave, while C_θ approached infinity ($C_\theta \to \infty$). On the other hand, as C_θ approached zero ($C_\theta \to 0$), the function became a sine wave.

In the flapping wing, viscous stress and pressure over the wing surface govern the aerodynamic forces. In the global coordinate system (x, y, z), the lift force, F_L is orthogonal to the stroke plane (x-z). The drag force, F_D is aligned in the y-direction, and is parallel to the stroke plane [3]. Therefore, the lift and drag forces can be decomposed into the coordinate forces; F_x, F_y, F_z. The lift is related to:

$$F_L = F_y \quad (8)$$

and the drag is:

$$F_D = F_x \cos\phi - F_z \sin\phi \quad (9)$$

The lift and drag coefficients are respectively:

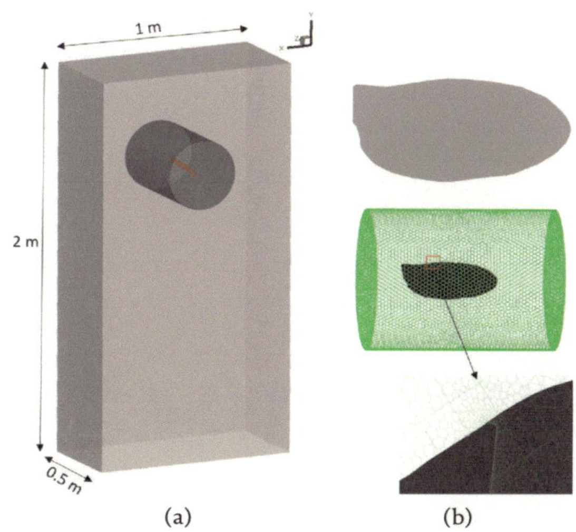

Figure 3.
(a) Computational setup with overset domain and symmetry walls, and (b) fruit fly wing planform used in the study with the meshed overset domain.

$$C_L = \frac{F_L}{0.5\rho U_0^2 S} \quad (10)$$

$$C_D = \frac{F_D}{0.5\rho U_0^2 S} \quad (11)$$

where S is the wing area. The reference velocity is defined as, $U_0 = 2\pi f R$, where R is the length of the wing.

The comparison of the lift and drag forces with the experimental data by Dickinson et al. [1] is shown in **Figure 4**, where $\tau = 0 - 0.5$ is downstroke and $\tau = 0.5 - 1$ is upstroke. The percent error obtained for the average lift and drag forces during a cycle are 3.4 and 7.2%, respectively.

3.2 Bumblebee kinematics

The present study utilizes the kinematic data of bumblebees in forward flight adopted from Xiong and Sun [36]. The kinematic data including amplitude, angle of attacks during the downstroke and upstroke, stroke plane deviation, the body angle and other parameters of bumblebee were calculated for different forward flights. **Table 1** shows the kinematic parameters selected for 0 m/s (hovering), 1 m/s and 2.5 m/s. Based on the Dudley and Ellington [37], and Xiong and Sun [36], the duration of wing flip is 0.22 of the cycle time at stroke reversal.

During the wing motion, the stroke angle (ϕ) and angle of attack (α) with respect to wing flip are defined by a sinusoidal function:

$$\phi(t) = \overline{\phi} + \phi \sin 2\pi f t \quad (12)$$

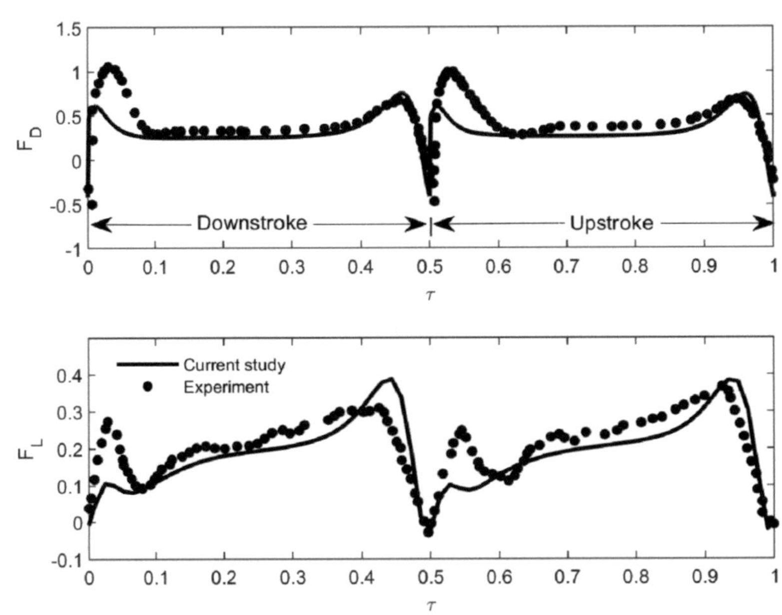

Figure 4.
Comparison of drag and lift forces with the experimental data by Dickinson et al. [1].

U (m/s)	*f (Hz)*	$\phi(^\circ)$	$\overline{\phi}(^\circ)$	$\beta(^\circ)$	$\alpha_d(^\circ)$	$\alpha_u(^\circ)$	$\chi(^\circ)$	*J*
0	155	116	1	6	27	21	46.8	0
1	150	115	16	21.6	28.5	24	31.8	0.12
2.5	150	115	23	28	24.5	34	25	0.31

Table 1.
Kinematic parameters of bee for different forward velocities adopted from ref. [36].

$$\alpha(t) = \begin{cases} \alpha_d + \frac{2\alpha_d}{\tau}\left(t - \frac{\tau}{2} - \frac{\tau}{2\pi}\sin\left(2\pi\frac{t-0.5\tau}{\tau}\right)\right), & 0 < t < \frac{\tau}{2} \\ \alpha_d, & \frac{\tau}{2} < t < t_1 \\ \alpha_d + \frac{\alpha_u - \alpha_d}{\tau}\left(t - t_1 - \frac{\tau}{2\pi}\sin\left(2\pi\frac{t-t_2}{\tau}\right)\right), & t_1 < t < t_2 \\ \alpha_u, & t_2 < t < t_3 \\ \alpha_u - \frac{2\alpha_u}{\tau}\left(t - t_3 - \frac{\tau}{2\pi}\sin\left(2\pi\frac{t-t_3}{\tau}\right)\right), & t_3 < t < 1 \end{cases} \tag{13}$$

where, $t_1 = 0.5(1-\tau)$, $t_2 = t_1 + \tau$ and $t_3 = 1 - 0.5\tau$, with the Reynolds number of:

$$Re = \frac{\rho U_0 R}{\mu} \tag{14}$$

where R is the length of wing ($R = 13.2$ mm) and μ is the dynamic viscosity of air. The maximum Reynolds number in the study is 7357.

3.3 Computational domain and overset mesh

The overset mesh, also named overlapping mesh, allows an analysis over multiple bodies that are moving or stationary. The method is an efficient way to both deal with the error related to moving boundaries and to circumvent the time-consuming complex mesh [38]. Moreover, it is easy to use, to sustain grid quality during the motion, and to avoid re-meshing malfunctions as well as generating simplified meshes for complex parts [39].

In the study, multiple overset mesh zones were used to simulate the wings and the body. The computational setup shown in **Figure 5(a)** has the dimensions of a $21R \times 17R \times 16R$ where the overset domain sizes of the wing and the body were $2.27R \times 1.81R \times 1.36R$ and $2.27R \times 1.51R \times 1.51R$, respectively (**Figure 5(b)**). The wing zone has a high-resolution grid than the body zone to capture the wake around the wing, as the background zone has coarser than the body zone.

This study considers two different models: thoracic and no thoracic models. In the thoracic model, the bee body is defined at a body temperature (T_b), with the bee transferring heat to the environment. The air properties change with ambient temperature (T_a) by specifying a polynomial relationship for the air dynamic viscosity. In the no thoracic model, the energy equation is not employed and therefore there is no heat transfer or dependence on temperature gradients. However, the fluid properties are specified using the ambient temperature for each case. The boundary conditions for the domain simulate a bee flying. The fluid velocity specified at the front of the

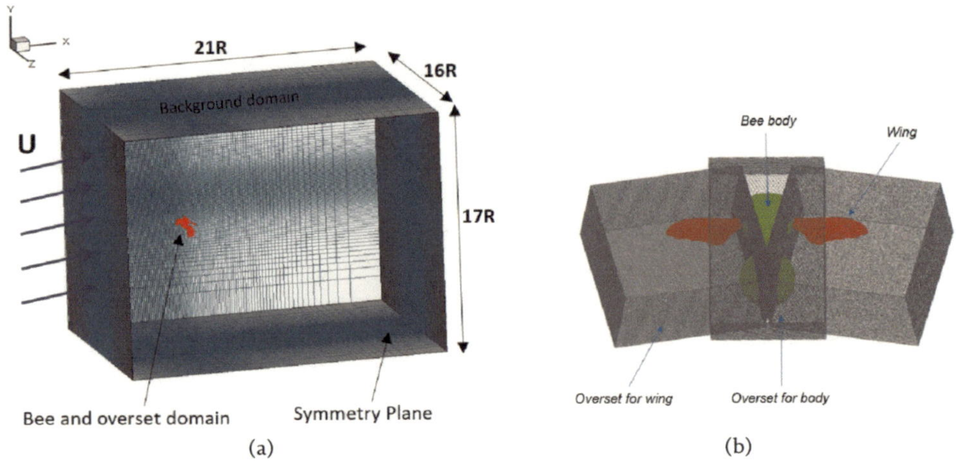

Figure 5.
(a) Computational domain with the bee and (b) overset mesh showing how the bee and wings are carefully resolved.

domain (inlet) represents the motion of the bee. The other five sides of the domain represent ambient conditions and therefore zero gage pressure is specified (outlet).

3.4 Grid Independence test

A systematic procedure to test grid independence is a vital procedure to determine the optimal grid size, thus guaranteeing that the accuracy of the numerical solution is independent of mesh size. In this study, the grid convergence index (GCI) methodology by Celik et al. [40] was used.

Three mesh samples named $coarse_3$, $medium_2$ and $fine_1$ were prepared for the analysis. As shown in **Table 2**, N represents the number of elements, and the grid refinement ratio r is the ratio of the average cell size height of two different grids. The grid convergence index GCI is defined as:

$$GCI = \frac{1.25}{r^p - 1}\left|\frac{f_1 - f_2}{f_1}\right| \tag{15}$$

where p is the order of accuracy and r is the grid refinement factor, defined as:

$$p = \frac{1}{\ln r}\ln\left|\frac{f_3 - f_2}{f_2 - f_1}\right| \tag{16}$$

$$r = h_2/h_1 \tag{17}$$

Grid	N	r	$\overline{C}_L$	GCI (%)
$Coarse_3$	1.2 M	$r_{32} = 1.312$	$\overline{C}_{L,3} = 1.2886$	$GCI_{32} = 0.867$
$Medium_2$	2.8 M	$r_{21} = 1.323$	$\overline{C}_{L,2} = 1.2708$	$GCI_{21} = 0.243$
$Fine_1$	6.4 M		$\overline{C}_{L,1} = 1.2766$	

Table 2.
GCI methodology parameters.

The variable h is the average mesh size and is calculated as:

$$h = \left[\frac{1}{N}\sum_{i=1}^{N}(\Delta V_i)\right]^{\frac{1}{3}} \tag{18}$$

where $\Delta\ V_i$ is the volume of the ith cell.

GCI analysis was performed by excluding the bee body; hence it only includes the total mesh count of the background and wing zone. To determine the GCI index, the average lift coefficient of a bumblebee hovering was considered. The $GCI_{32} = 0.867\%$ for the coarse-medium comparison grids and $GCI_{21} = 0.243\%$ for the medium-fine comparison (see **Table 2**). The lift coefficient of these three meshes for a cycle is shown in **Figure 6** and the differences are minimal. Hence, the medium grid is chosen as the base mesh for subsequent simulations.

4. Results and discussions

The aerodynamic behavior of bumblebees for different environmental conditions was analyzed using a morphologically accurate bumblebee model and computational fluid dynamics. The effects of heat transfer between bee body and environment, and ambient conditions were analyzed for different temperature levels at three different forward speeds. The findings of the study are detailed with average lift-thrust forces, aerodynamic power, the time history of the forces and the power for a cycle and the visualization of temperature, pressure, and iso-surfaces with Q-criterion.

The flapping wing motion of the bee for a cycle is illustrated in **Figure 7**, where $\tau = 0 - 0.5$ refers to downstroke and $\tau = 0.5 - 1$ represents upstroke. During downstroke, the wing moves forward while it moves backward during upstroke in relation to the bee's longitudinal axis. At the end of each stroke, the wing performs a rapid rotation and moves in the reverse direction.

The transient simulations were performed for 5 cycles. **Figure 8** shows the time history of lift (F_L) and thrust (F_T) forces of the thoracic model at 1 m/s and $T_b = 32°C$ and $T_a = 20°C$ for five cycles. The figure for the no thoracic model is virtually the same as the thoracic model, therefore only the thoracic model is shown in **Figure 8**. Even though there are no major differences in the cycles, in order to ensure transient effects are minimized, the third cycle ($\tau = 0 - 1$) was chosen as the data used in the tables and figures for the subsequent simulations.

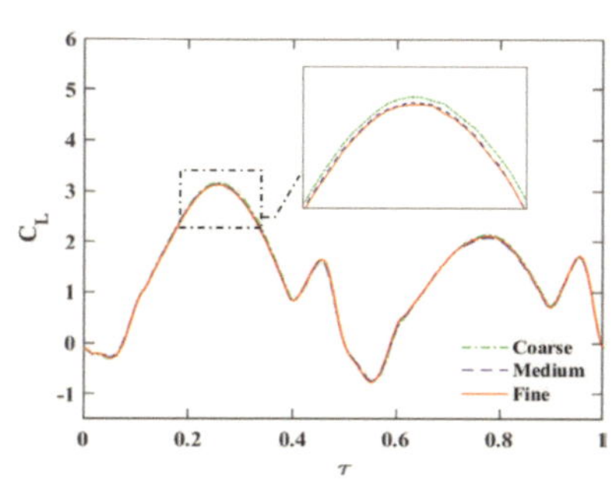

Figure 6.
Lift coefficient of the wing for different meshes at hovering position.

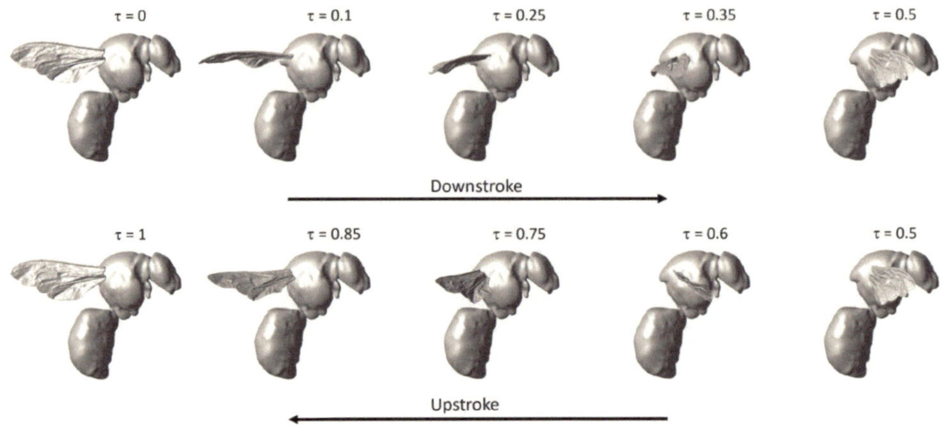

Figure 7.
Kinematics of the flapping wing of the bee for a cycle, where $\tau = 0 - 0.5$ is downstroke and $\tau = 0.5 - 1$ is upstroke.

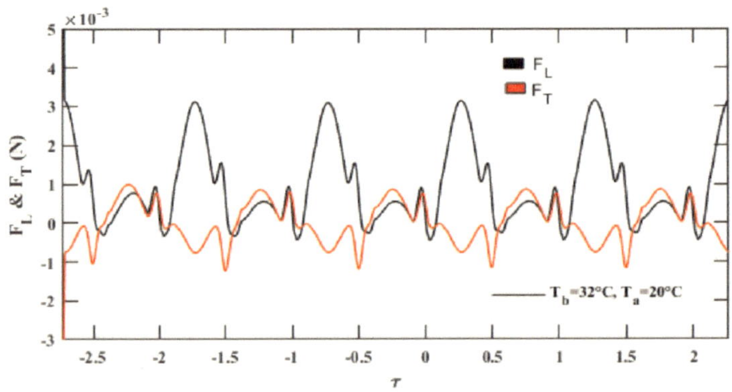

Figure 8.
Time history of lift and thrust forces at 1 m/s and $T_b = 32°C$ and $T_a = 20°C$ for five cycles.

The time history of lift and thrust forces of no thoracic model for a cycle is shown in **Figure 9** at 0, 1, and 2.5 m/s for the different ambient temperatures (T_a). The downstroke is the most valuable part of lift production due to a larger leading-edge vortex (LEV) formation. At forward velocities, the majority of lift is produced during downstroke, but at hovering is produced during downstroke and upstroke. Bee flying at hovering produces less lift and thrust forces than forward speeds due to the absence of forward velocity.

The thrust production mainly occurs during the upstroke and increases with increasing forward velocity. It is observed that during the downstroke, the negative thrust (drag) has a positive correlation, but it decreases during stroke reversal ($\tau = 0.4 - 0.6$) with increasing forward flight. As shown in **Figure 9**, the increasing ambient temperature causes a drop in the lift and thrust at all speed levels. The majority of the lift loss takes place during the downstroke, where the main part of the lift is produced. In the upstroke, there is a decrease in the lift with increasing temperature, but it is not an apparent difference except in the hovering model. The thrust also drops with ambient temperature changes in the upstroke. However, the drag in the upstroke decreases with increasing temperature, which is higher at the mid-downstroke in the hover position.

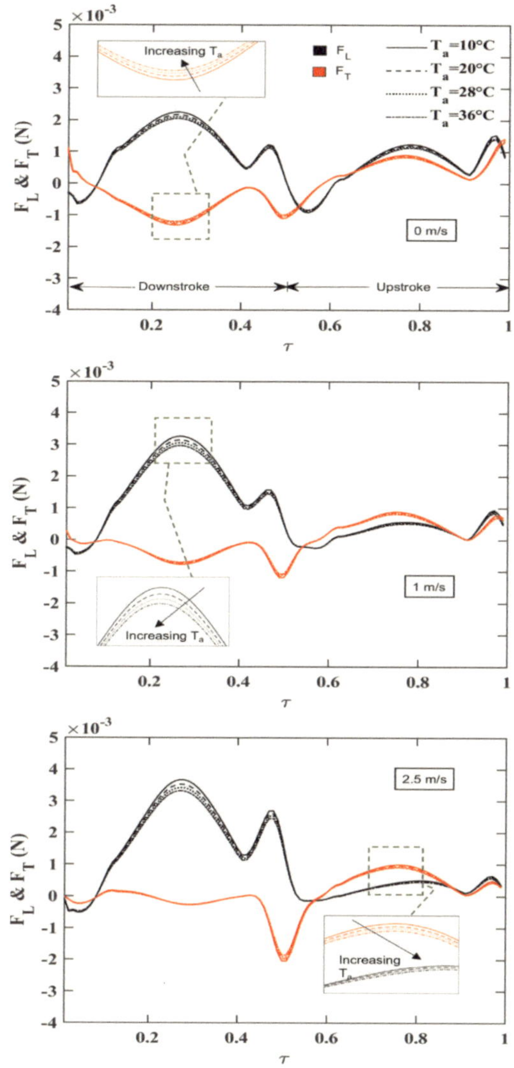

Figure 9.
Time history of lift and thrust forces for a cycle at 0 m/s (hovering), 1 and 2.5 m/s for different ambient temperatures.

The average lift and thrust forces are shown in **Table 3** for thoracic and no thoracic models. The base cases assume negligible heat transfer, while two models at every velocity level are considered for the thoracic activity with ($T_b = 32°C$) and ($T_b = 38°C$) varying ambient temperatures (T_a). Increasing forward velocity increases the average lift and thrust forces in all conditions; thus, flying at 2.5 m/s has higher lift and thrust than 1 and 0 m/s. The bee experiences negative thrust (drag) force while hovering due to the absence of forward velocity. It is evident from the table that the average F_L and F_T for the models with $T_b = 32°C$ and $T_a = 20°C$, and $T_b = 38°C$ and $T_a = 28°C$ show minimal changes compared to the models without a thoracic region, thus indicating the limited impact of heat transfer due to the bee's high frequency flapping to allow density changes.

	Temperature (°C)	0 m/s	1 m/s	2.5 m/s
$\overline{F}_L$ ($\times 10^4 N$)	$T_a = 10$	8.242	9.611	10.907
	$T_a = 20$	7.945	9.260	10.491
	$T_a = 20$, $T_b = 32$	7.943	9.250	10.478
	$T_a = 28$	7.718	9.001	10.194
	$T_a = 28$, $T_b = 38$	—	8.970	10.157
	$T_a = 36$	7.499	8.692	9.826
$\overline{F}_T$ ($\times 10^4 N$)	$T_a = 10$	−0.937	0.0181	0.637
	$T_a = 20$	−0.901	0.0165	0.601
	$T_a = 20$, $T_b = 32$	−0.903	0.0148	0.596
	$T_a = 28$	−0.873	0.0162	0.581
	$T_a = 28$, $T_b = 38$	—	0.0153	0.573
	$T_a = 36$	−0.846	0.0154	0.541

Table 3.
Average lift and thrust forces for different environmental conditions at 0, 1, and 2.5 m/s.

In contrast to thoracic activity, the effect of ambient temperature is prevalent. The higher ambient temperatures result in reduced lift and thrust forces. When T_a increases from 10°C to 36°C, the lift decreases by 9% for hovering, 9.6% for 1 m/s and 10% for 2.5 m/s, revealing that the bee experiences more lift loss flying at higher speeds. Moreover, the thrust in the hover position falls by 14.97 and 15% for 1 and for 2.5 m/s, respectively. Contrary to forward flight, the bee experiences negative thrust (drag) in hovering due to the lack of forward velocity. The drag during hovering decreases by 9.71% when the ambient temperature increases from 10°C to 36°C. As a result, warmer environments negatively affect the bee overall performance, while the effect of the thoracic activity is limited.

The temperature contours of the thoracic model ($T_b = 32°C, T_a = 20°C$) are shown in **Figure 10** at 2.5 m/s. **Figure 10(a)** shows temperature changes around the bee. The heat plume from the body alters the flow conditions by increasing the air temperature around the bee, resulting in density reduction and viscosity increase of air. The distribution of temperature over the wing dorsal surface is shown in **Figure 10(b)**. The heat plume emitted from the body increases the temperature on the forewing and hindwing. Specifically, the temperature on the leading-edge and upper part of the hindwing regions are similar, as the temperature increase on the trailing-edge of the hindwing is higher. As mentioned above, due to the limited effect of the thoracic activity, the remainder of the discussion and results are for the "no thoracic model" in which only ambient temperature is changed.

Figure 11 depicts the suction pressure (negative pressure because it is gage pressure) along the dorsal side of the wing during the downstroke at a time instant of $\tau = 0.25$ for the lowest (10°C) and highest (36°C) ambient temperatures at 0, 1, and 2.5 m/s. The bee at 2.5 m/s has lower suction pressure on the dorsal due to high spanwise velocity than 0 m/s and 1 m/s, leading to larger lift production. The results show that there are changes in the pressure distribution on the forewing and hindwing as the ambient temperature increases from 10°C to 36°C, resulting in increasing suction pressure at all speed levels, resulting in decreasing the lift. Particularly, the

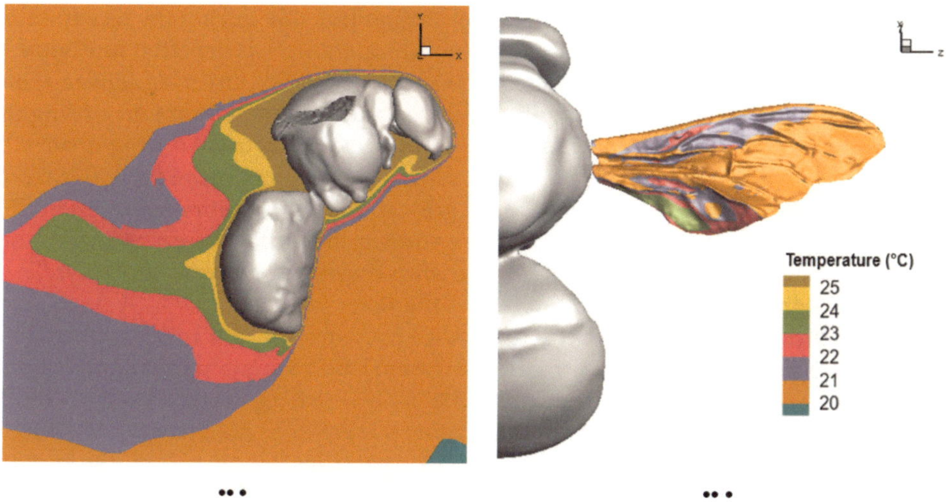

Figure 10.
Temperature contours showing the flow conditions around the bee (a) and the temperature distribution on the wing dorsal side (b) at 2.5 m/s ($\tau = 0.33$) for $T_b = 32°C$, $T_a = 20°C$.

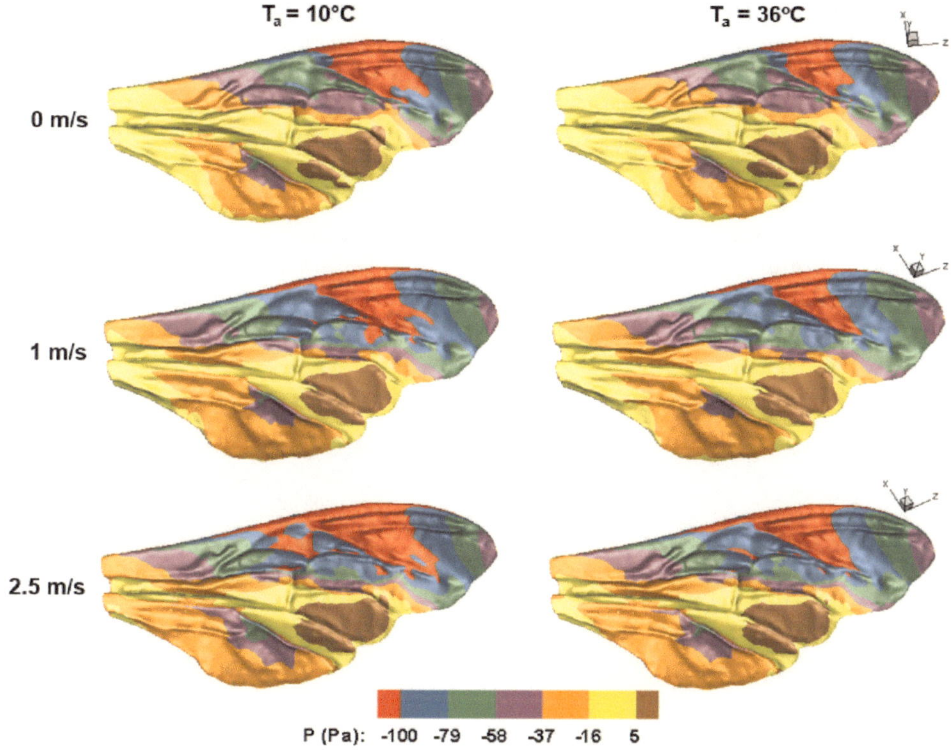

Figure 11.
Pressure (gage) contours on the dorsal side of the wing for different temperature and velocity levels at mid-downstroke ($\tau = 0.25$).

low-pressure regions on the leading-edge, wingtip, and hindwing are reduced, resulting in a decrease in the lift force. Furthermore, there are pressure increases spotted on the root side of the wing for 1 and 2.5 m/s, which also contribute to lift reduction.

Additionally, the pressure distribution on the ventral side during the upstroke at $\tau = 0.75$ for the lowest and the highest ambient temperatures is depicted in **Figure 12**. The figure only shows the extreme cases of 0 and 2.5 m/s. For hovering, lower pressure is experienced on the wing tip of the ventral, leading to more lift during the upstroke, proving that more lift production takes place at hovering during upstroke and downstroke than forward flight. When the ambient temperature increases from $10^{\circ}C$ to $36^{\circ}C$, the differences in pressure are greater at 2.5 m/s compared to 0 m/s. The reduction in the size of the negative pressure zone in the wingtip area, as a result of increased ambient temperature, is associated with decreased lift and thrust, while there are no significant changes in the rest of the wing surface.

Figure 13 displays the vortex structures with pressure contour following stroke reversal using the Q-criterion at $\tau = 0.33$ for 1 m/s. LEV formation during the downstroke makes an important contribution to the lift by creating suction pressure on the dorsal side of the wing. Increasing ambient temperature does not change LEV but small changes in the trailing edge vortex (TEV) structures. Furthermore, the suction pressure on the leading-edge and the hindwing increases with increasing ambient temperature, thereby leading to reducing lift.

The iso-surfaces with pressure (gage) contours for the lowest and the highest temperatures at 0 and 2.5 m/s are presented in **Figure 14**. At higher velocities, a thicker tip vortex (TV) is generated due to the availability of strong spanwise flow, so the 2.5 m/s model predicts lower pressure over the wing than 0 m/s. The trailing-edge vortex (TEV) structures at 0 m/s are larger than at 2.5 m/s, leading to higher lift and drag. When ambient temperature increases from $10^{\circ}C$ to $36^{\circ}C$, vortex structures do not change considerably for 0 and 2.5 m/s. On the wing leading-edge and hindwing, while there is no significant increase in pressure at 0 m/s, the suction pressure at 2.5 m/s is considerably affected by increasing ambient temperature, resulting in more lift drop. The Q-criterion iso-surfaces with vorticity magnitude for the lowest and highest temperatures during upstroke are shown in **Figure 15** at 2.5 m/s. Small changes are observed in iso-surfaces and vorticity magnitude with increasing temperature that result in slight differences in lift and thrust at $\tau = 0.65$.

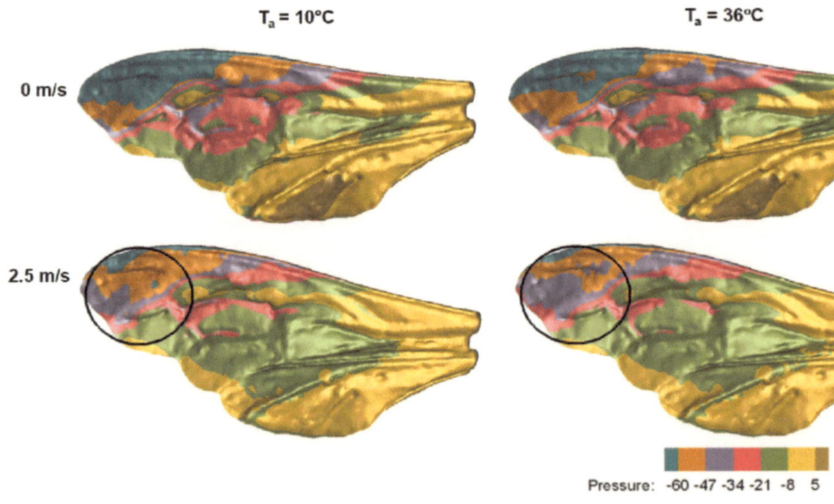

Figure 12.
Pressure (gage) contours on the ventral side of the wing for different temperature and velocity levels at 0 and 2.5 m/s at mid-upstroke ($\tau = 0.75$).

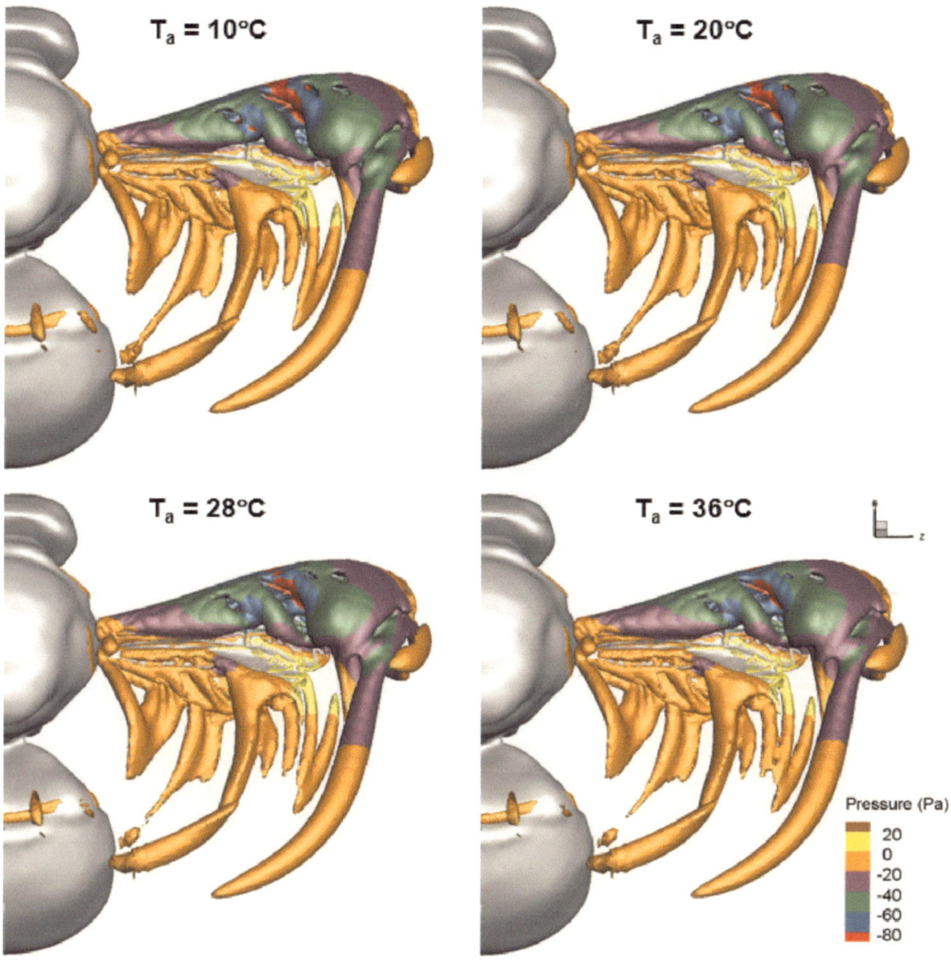

Figure 13.
Iso-surfaces with Q-criterion and pressure distribution (5×10^6) at 1 m/s ($\tau = 0.33$).

Figure 16 illustrates the iso-surfaces with vorticity magnitude at 2.5 m/s and different time periods (τ) for the lowest and highest ambient temperatures. Note that there are no major changes in the vorticity structures and magnitudes except at $\tau = 0.55$, in which iso-surfaces at the wingtip disappeared when the ambient temperature increased from $10^{\circ}C$ to $36^{\circ}C$.

Bees alter their wing beat frequency when they need to increase or decrease lift and thrust forces. Under the same kinematic parameters, a bee flying in warmer environments must change its flapping frequency to maintain flight performance. When ambient temperature increases from $10^{\circ}C$ to $36^{\circ}C$, the bee must increase its flapping frequency to compensate for the lift loss. The average lift forces of the bee at the lowest and highest temperatures for three different velocities are shown in **Table 4** with the adjusted frequencies (f_{adj}) used in the simulations. Using the lift forces, an interpolation was performed to determine the adjusted frequencies of the bee flying at $36^{\circ}C$. Then these frequencies were validated by performing the new simulations, which proved that the lift lost at $36^{\circ}C$ was regained using the adjusted

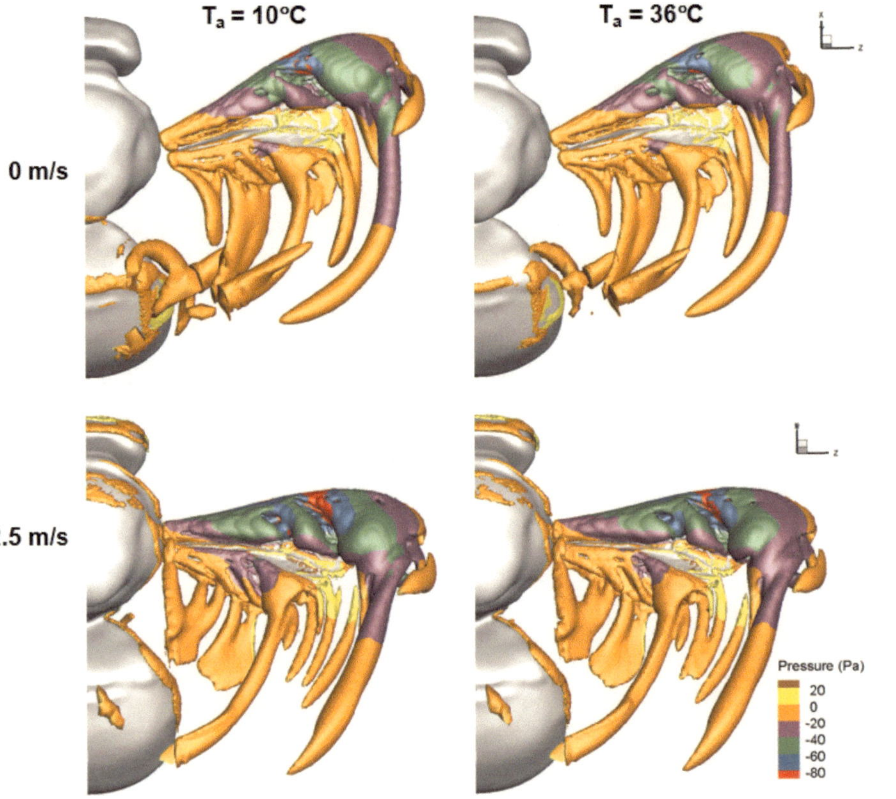

Figure 14.
Iso-surfaces with Q-criterion and pressure distribution (3×10^6) at 0 and 2.5 m/s ($\tau = 0.33$).

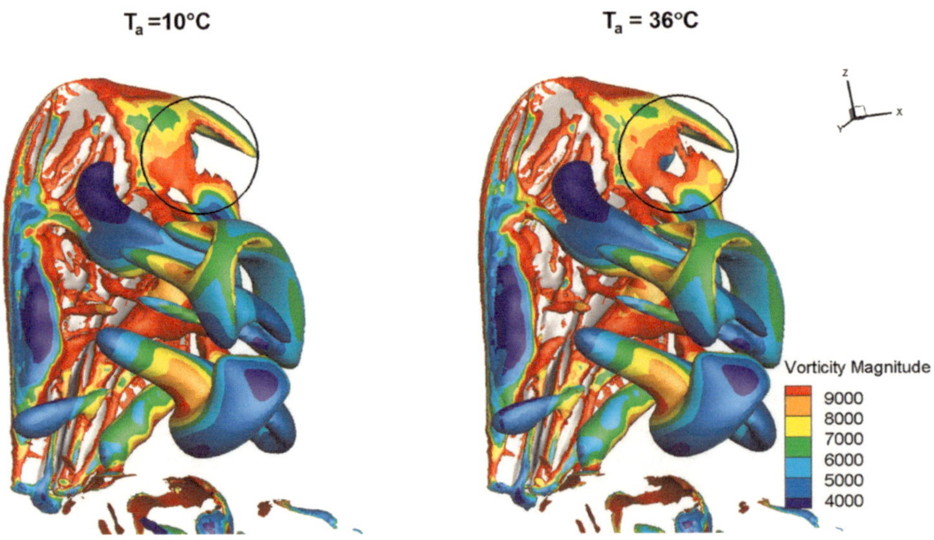

Figure 15.
Iso-surfaces with Q-criterion and vorticity magnitude (3×10^6) at 1 m/s ($\tau = 0.65$).

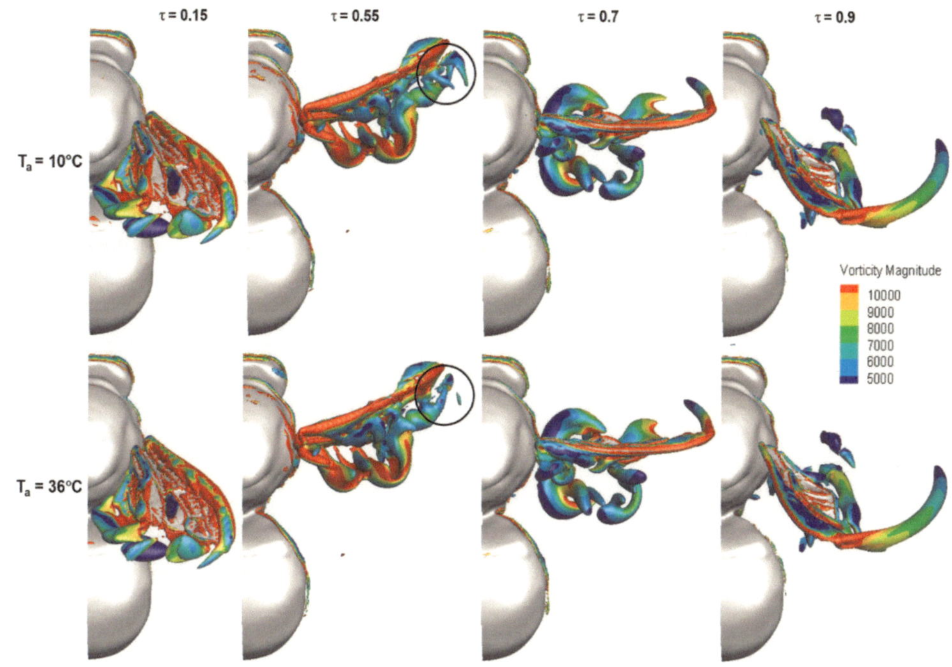

Figure 16.
Iso-surfaces with Q-criterion and vorticity magnitude (4×10^6) at 2.5 m/s for different time periods in a cycle.

	Temperature	0 m/s	1 m/s	2.5 m/s
$\overline{F}_L$ ($\times 10^4$ N)	$T_a = 10^\circ C$	8.242	9.611	10.907
		($f = 155$ Hz)	($f = 150$ Hz)	($f = 150$ Hz)
	$T_a = 36^\circ C$	7.499	8.692	9.826
		($f = 155$ Hz)	($f = 150$ Hz)	($f = 150$ Hz)
	$T_a = 36^\circ C$	8.208	9.606	10.937
		($f_{adj} = 162$ Hz)	($f_{adj} = 159$ Hz)	($f_{adj} = 158$ Hz)

Table 4.
Average lift forces with the adjusted frequency of the lowest (10°C) and the highest ambient temperature (36°C) models at 0, 1, and 2.5 m/s.

frequencies. The difference in the lift between the original and adjusted frequency is 0.4% for 0 m/s, 0.05% for 1 m/s and 0.2% for 2.5 m/s. To maintain the lift requirement flying at $36^\circ C$ ambient, the bee must increase the flapping frequency from 155 to 162 Hz for 0 m/s, 150 to 158 Hz for 1 m/s, and 150 to 159 Hz for 2.5 m/s. The difference in the flapping frequency for $10^\circ C$ and $36^\circ C$ is 4.5% for 0 m/s, 6% for 1 m/s, and 5.3% for 2.5 m/s.

Aerodynamic power for flapping flight represents the power to overcome the fluid forces, defined as $P = M\omega$, where M and ω are the moment and the angular velocity of the wing. The time history of the aerodynamic power of the bee at different forward speeds is shown in **Figure 17** for the lowest and highest ambient temperature with the original and adjusted frequencies. The majority of power production takes place during the downstroke and increasing forward speed increases the aerodynamic power; therefore, the 2.5 m/s model generates more power during the downstroke. Moreover, the power reduction during stroke reversal ($\tau = 0.4 - 0.6$) is maximum at 2.5 m/s and minimum at 1 m/s, indicating a higher average power generation of 1 m/s than 0 m/s and 2.5 m/s. In the upstroke, the power generation at 0 m/s is higher than at 1 m/s but lower than at 2.5 m/s. A larger decline in power at 0 m/s was also observed at the end of the upstroke. Increasing temperature decreases the power production during downstroke and upstroke at all forward speeds but is maximum at

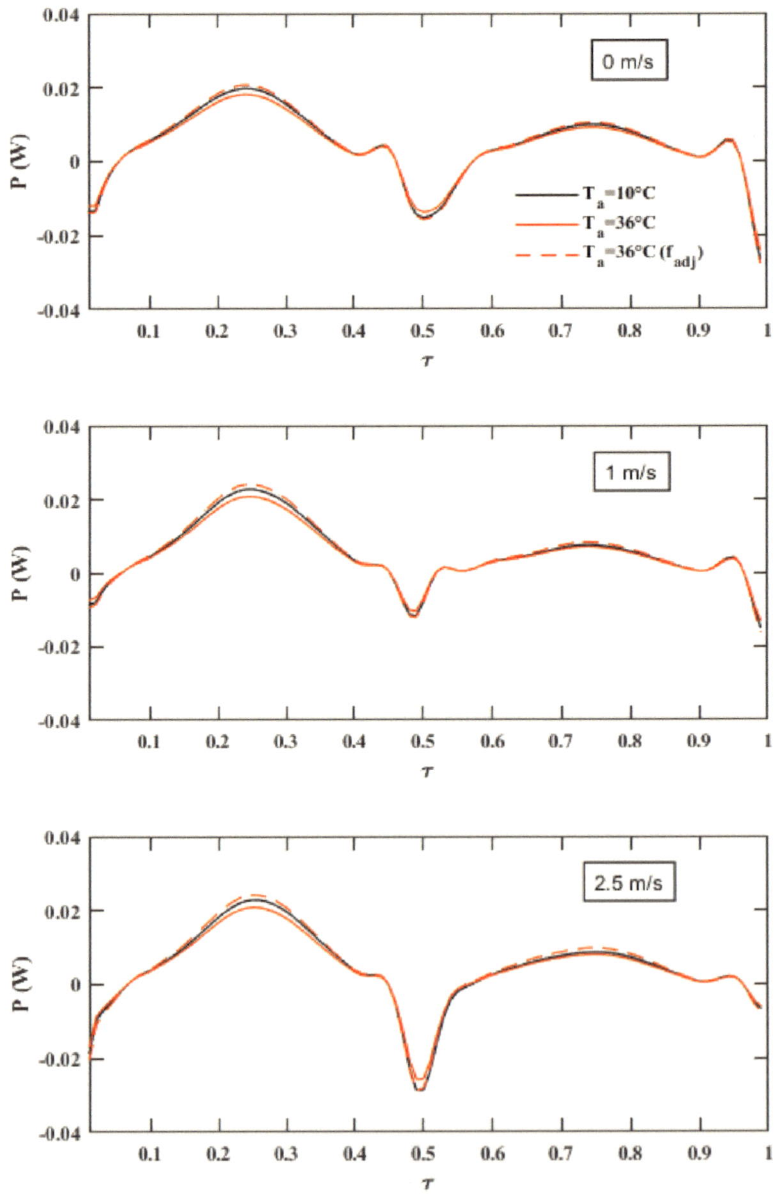

Figure 17.
Time history of aerodynamic power of 10°C and 36°C models for one cycle at 0, 1, and 2.5 m/s for the original and adjusted frequencies.

	Temperature	0 m/s	1 m/s	2.5 m/s
$\overline{P}$ (mW)	$T_a = 10°C$ ($f_{orig.}$)	3.8	5.0	4.1
	$T_a = 36°C$ ($f_{orig.}$)	3.5	4.6	3.8
	$T_a = 36°C$ ($f_{adj.}$)	4.0	5.3	4.6

Table 5.
Average aerodynamic power of 10°C and 36°C models at 0, 1, and 2.5 m/s for the original and adjusted frequencies.

mid-downstroke. The adjusted wing-beat frequencies shown in **Table 4** compensate for the power at all speed levels that was lost with the ambient temperature rise from 10°*C* to 36°*C*. The power increase is mostly observed during the downstroke and upstroke, and it is maximum at mid-downstroke.

The average aerodynamic power of the lowest and highest temperature models with the original and adjusted frequency model is illustrated in **Table 5**. The average aerodynamic power of the bee at 1 m/s is higher than at 0 and 2.5 m/s. When ambient temperature increases from 10°*C* to 36°*C*, the aerodynamic power drops by 7.9% for 0 m/s, 8% for 1 m/s and 7.3% for 2.5 m/s, indicating that the bee suffers more aerodynamic power loss at low speeds. The bee employs the adjusted frequencies to compensate for the loss. The difference in the average aerodynamic power between 10°*C* model and the adjusted frequency model of 36°*C* is 5% for 0 m/s, 6% for 1 m/s, and 12% for 2.5 m/s. Thus, the adjusted frequencies require more power for the 2.5 m/s model.

5. Conclusion

This study investigated the impacts of heat transfer and ambient temperature changes on bee flight performance, highlighting a better understanding of the roles of thoracic and ambient temperature fluctuations on bumblebee aerodynamic performance. Heat transfer resulting from thoracic temperature activity has a rather limited impact on bee performance as could be detected by the rigid-wing models. It was observed that heat transfer with the environment increases the temperature of the wing surface and changes the flow conditions around the bee. Hence, the lift and thrust forces of the thoracic model decreased when compared to the no thoracic model for the same ambient condition, indicating that thoracic activity has an impact, however limited.

Ambient temperature change has more impact on the aerodynamics of the bee than thoracic temperature activity. Increasing ambient temperature changes the flow properties, thereby resulting in lower lift. Thrust force reduces at 1 m/s and 2.5 m/s, as the drag (negative thrust at hovering) becomes lower at 0 m/s (hovering) with increasing ambient temperature. Increasing temperature from the lowest temperature (10°*C*) to the highest temperature (36°*C*) caused 9, 9.6, and 10% less lift at 0, 1, and 2.5 m/s, respectively, underlining more lift loss at higher velocities. The thrust dropped 14.97 and 15% at 1 and 2.5 m/s; on the contrary, the drag decreased at hovering by 9.71%. Furthermore, the aerodynamic power decreased by 7.9% for 0 m/s, 8% for 1 m/s and 7.3% for 2.5 m/s when ambient temperature increased from

10°*C* to 36°*C*. To compensate for the lift loss, the adjusted frequency was specified. The lift losses could be recovered, while increasing the aerodynamic power by 5% for 0 m/s, 6% for 1 m/s and 12% for 2.5 m/s.

As a result, warmer temperatures reduce bee performance by negatively affecting thoracic temperature and flow condition variations around the bees. Heat transfer resulting from thoracic and environmental temperature changes, reduces the aerodynamic performance of bees. While the environmental effects in bee performance can be clearly determined, the real influence of thoracic temperature change cannot be accurately analyzed without improving the computational model. For example, laboratory experiments that can measure the flow physics when a bee is subjected to different environmental temperatures can provide data to verify the models and guide model improvements. Overall, it can be deduced that bees are able to compensate loss of lift and power due to increasing temperatures with a rise in their flapping frequency. However, it is envisaged that rising environmental heat, requiring higher wing beats, will result in increased thoracic temperature. There comes a time that the combined temperature effects supersede the bee's muscular abilities to compensate for lift and power, hence, drastically reducing the bee's endurance or limiting its foraging perimeter.

Author details

Isa Gan, Munjal Shah, Javid Bayandor and Francine Battaglia*
The Department of Mechanical and Aerospace Engineering, University at Buffalo, The State University of New York, Buffalo, NY, USA

*Address all correspondence to: fbattagl@buffalo.edu

References

[1] Dickinson MH, Lehmann FO, Sane SP. Wing rotation and the aerodynamic basis of insect flight. Science. 1999;**284**(5422):1954-1960

[2] Nagai H, Isogai K, Fujimoto T, Hayase T. Experimental and numerical study of forward flight aerodynamics of insect flapping wing. AIAA Journal. 2009;**47**(3):730-742

[3] Lyu YZ, Zhu HJ, Sun M. Aerodynamic forces and vortical structures of a flapping wing at very low Reynolds numbers. Physics of Fluids. 2019;**31**(4):041901

[4] Feaster J, Battaglia F, Bayandor J. A computational study on the influence of insect wing geometry on bee flight mechanics. Biology Open. 2017;**6**: 1784-1795

[5] Shah M, Battaglia F, Bayandor J. The importance of morphology in further unraveling the bumblebee flight paradox. Journal of Fluids Engineering. 2023;**145**(1):011303

[6] Shah M, Bayandor J, Battaglia F. The effect of wing corrugation on insect forward flight. In: Proceeding of 5-6th Thermal and Fluids Engineering Conference (TFEC). Virtual: Begellhouse; 2021. p. 636

[7] Shyy W, Lian Y, Tang J, Viieru D, Liu H. Aerodynamics of Low Reynolds Number Flyers. Cambridge Aerospace Series. England, United Kingdom: Cambridge University Press; 2007

[8] Tobing S, Young J, Lai JCS. Effects of wing flexibility on bumblebee propulsion. Journal of Fluids and Structures. 2017;**68**:141-157

[9] Gan I, Shah M, Bayandor J, Battaglia F. The effect of thermoregulation on the aerodynamics of bumblebees. In: Proceeding of 8th Thermal and Fluids Engineering Conference (TFEC). College Park, USA: Begellhouse; 2023. pp. 171-174

[10] Heinrich B. The physiology of exercise in the bumblebee: Bumblebees are limited in their superb athletic performance by extremes of air temperature, but by regulating their body temperature they achieve considerable control. American Scientist. 1977;**65**(4):455-465

[11] Heinrich B. Temperature regulation in the bumblebee *Bombus vagans*: A field study. Science. 1972;**175**(4018):185-187. Number: 4018 Publisher: American Association for the Advancement of Science

[12] Heinrich B, Kammer AE. Activation of the fibrillar muscles in the bumblebee during warm-up, stabilization of thoracic temperature and flight. Journal of Experimental Biology. 1973;**58**(3): 677-688

[13] Roberts SP, Harrison JF. Mechanisms of thermoregulation in flying bees. American Zoologist. 1998;**38**(3):492-502

[14] Heinrich B, Esch H. Thermoregulation in bees. American Scientist. 1994;**82**:164-170

[15] Kenna D, Pawar S, Gill RJ. Thermal flight performance reveals impact of warming on bumblebee foraging potential. Functional Ecology. 2021;**35**: 2508-2522

[16] Nixon AE, Hines H. Bumblebee thermoregulation: Understanding the thermal properties of physical features of bumblebees. The Penn State McNair Journal. 2017;**2017**:99

[17] Heinrich B, Heinrich MJE. Size and caste in temperature regulation by bumblebees. Physiological Zoology. 1983;**56**(4):552-562

[18] Heinrich B. Bumblebee Economics. Cambridge, Massachusetts: Harvard University Press; 2004

[19] Ellington CP. The aerodynamics of hovering insect flight. II. Morphological parameters. Philosophical Transactions of the Royal Society of London B, Biological Sciences. 1984;**305**(1122):17-40

[20] Norberg UM. In: Bradshaw SD, Burggren W, Heller HC, Ishii S, Langer H, Neuweiler G, et al., editors. Vertebrate Flight. Vol. 27 of Zoophysiology. Berlin, Heidelberg: Springer Berlin Heidelberg; 1990

[21] Chin DD, Lentink D. Flapping wing aerodynamics: From insects to vertebrates. Journal of Experimental Biology. 2016;**219**(7):920-932

[22] Shyy W, Berg M, Ljungqvist D. Flapping and flexible wings for biological and micro air vehicles. Progress in Aerospace Sciences. 1999; **35**:51

[23] Wang ZJ. Aerodynamic efficiency of flapping flight: Analysis of a two-stroke model. Journal of Experimental Biology. 2008;**211**(2):234-238

[24] Gérard M, Martinet B, Dehon M, Rasmont P, Williams PH, Michez D. The utility of wing morphometrics for assigning type specimens to cryptic bumblebee species. Systematic Entomology. 2020;**45**:849-856

[25] Shyy W, Ck K, Chirarattananon P, Ravi S, Liu H. Aerodynamics, sensing and control of insect-scale flapping-wing flight. Proceedings of the Royal Society A. 2016;**472**(2186):20150712

[26] Shah M, Bayandor J, Battaglia F. Compensating bumblebee flight aerodynamics due to wing wear. In: Proceeding of 7th Thermal and Fluids Engineering Conference (TFEC). Las Vegas, NV, USA: Begellhouse; 2022. pp. 1299-1302

[27] Church NS. Heat loss and the body temperatures of flying insects II. Heat conduction within the body and its loss by radiation and convection. The Journal of Experimental Biology. 1960;**37**: 186-212

[28] Woods WA, Heinrich B, Stevenson RD. Honeybee flight metabolic rate: Does it depend upon air temperature? Journal of Experimental Biology. 2005;**208**(6):1161-1173. Number: 6

[29] Heinrich B. Thermoregulation in endothermic insects: Body temperature is closely attuned to activity and energy supplies. Science. 1974;**185**(4153): 747-756

[30] Esch H, Goller F, Heinrich B. How do bees shiver? The Science of Nature. 1991;**78**:7. 1991;78(7):325–328. Number: 7 Publisher: Springer

[31] Chappell MA. Temperature regulation of carpenter bees (*Xylocopa californica*) foraging in the Colorado Desert of Southern California. Physiological Zoology. 1982;**55**(3): 267-280

[32] Church NS. Heat loss and the body temperatures of flying insects I. Heat loss by evaporation of water from the body. The Journal of Experimental Biology. 1960;**37**:171-185

[33] Heinrich B. Heat exchange in relation to blood flow between thorax and abdomen in bumblebees. The

Journal of Experimental Biology. 1976; **64**(3):561-585

[34] Harrison JF, Fewell JH, Roberts SP, Hall HG. Achievement of thermal stability by varying metabolic heat production in flying honeybees. Science. 1996;**274**(5284):88-90

[35] Vogt FD. Thermoregulation in bumblebee colonies. I. Thermoregulatory versus brood-maintenance behaviors during acute changes in ambient temperature. Physiological Zoology. 1986;**59**(1):55-59

[36] Xiong Y, Sun M. Dynamic flight stability of a bumblebee in forward flight. Acta Mechanica Sinica. 2008; **24**(1):25-36

[37] Dudley R, Ellington CP. Mechanics of forward flight in bumblebees: I. Kinematics and morphology. Journal of Experimental Biology. 1990;**148**(1):19-52

[38] Ma ZH, Qian L, Martínez-Ferrer PJ, Causon DM, Mingham CG, Bai W. An overset mesh based multiphase flow solver for water entry problems. Computers & Fluids. 2018;**172**:689-705

[39] Ramakrishnan S, Scheidegger T. Overset meshing in ANSYS fluent. In: Proceedings of the 13th Symposium on Overset Composite Grids and Solution Technology, Mukilteo, WA, USA. 2016

[40] Celik IB, Ghia U, Roache PJ, Freitas CJ. Procedure for estimation and reporting of uncertainty due to discretization in CFD applications. Journal of Fluids Engineering. 2008; **130**(7):078001